A NEOTROPICAL COMPANION

A
Neotropical
Companion

An Introduction to the
Animals, Plants, and
Ecosystems of the
New World Tropics

BY JOHN C. KRICHER

ILLUSTRATED BY
ANDREA S. LEJEUNE

PRINCETON UNIVERSITY

PRESS

Copyright © 1989 by Princeton University Press
Published by Princeton University Press, 41 William Street,
Princeton, New Jersey 08540
In the United Kingdom: Princeton University Press,
Chichester, West Sussex

Library of Congress Cataloging-in-Publication Data

Kricher, John C.
   A neotropical companion : an introduction to the
animals, plants, and ecosystems of the New World tropics /
John C. Kricher ; illustrated by Andrea S. LeJeune.
   p. cm.
Bibliography: p.
Includes index.
ISBN 0-691-08520-X (alk. paper) ISBN 0-691-08521-8 (pbk.)
1. Ecology—Latin America. 2. Natural history—Latin
America.
I. Title.
QH106.5.K75 1989
574.5'2642'098—dc19   88-21483

This book has been composed in Linotron Baskerville

Princeton University Press books are printed on acid-free
paper, and meet the guidelines for permanence and
durability of the Committee on Production Guidelines for
Book Longevity of the Council on Library Resources

Printed in the United States of America

16   15   14   13   12   11   10   9   8   7   6   5

To my mother,
and to memories of my father

CONTENTS

PREFACE

THIS IS a book about the American tropics, the lands of Central and South America, their great rain forests, and the creatures that dwell within them. The New World tropics, or Neotropics, provide some of the most remarkable examples of natural history known. In recent years, many North Americans have ventured to the tropics in search of exotic birds, mammals, insects, and plants and to see firsthand the amazing rain forests. Students are also experiencing the American tropics as more and more college courses focus on Latin America and include trips to tropical field stations. Scientific knowledge about the American tropics has burgeoned dramatically. Research performed at the Smithsonian Institution field station on Barro Colorado Island in the Canal Zone of Panama, at the various field stations of the Organization for Tropical Studies in Costa Rica, as well as at many other places has demonstrated that the tropics abound with extraordinary examples of nature's complexity. Many, if not most, of these examples have appeared thus far scattered in the technical literature and have yet to find their way to a general book. This book attempts to remedy that situation.

The basis for this book developed from an undergraduate course I teach at Wheaton College as well as from my Chautauqua short course "Ecology and Evolution of the Tropics." I have relied heavily on my own experiences in Central and South America, especially Belize, a delightful little country on the southern Yucatán Peninsula. Research is a human endeavor, and, thus, in discussing the many fascinating aspects of tropical ecology and natural history, I have included the names of many (but by no means all) of the people who are in the field, performing the research that is yielding the information. Though this is not meant as a highly technical book, I include a references section so that the interested reader can read firsthand any of the studies I have used in writing this book. Unfortunately, but by necessity, I have had to be highly selective, and many excellent studies have been omitted. I believe, however, that those included will serve well to introduce you to what it is about the American tropics that makes me and my many colleagues want to keep returning and studying. For the reader with little or no previous knowledge or formal preparation in ecology, I have included a glossary of ecological terms, especially oriented to tropical ecology. I have also defined terms in the text, when the term is first used.

The book begins with an overview of the tropical climate, the importance of seasonality, and the various kinds of habitats found in the American tropics. From there I discuss both the structure and function of the rain forests, including methods by which native Americans farm the tropics. I then focus on evolution in the tropics, on why the tropics host so many myriads of species, and how species evolve and interact. A major emphasis in this book is the topic of coevolution, when two or more species interact in particularly intricate and interdependent ways. A chapter titled "The Tropical Pharmacy" discusses the many drugs present in tropical vegetation and the evolutionary influence of these drugs. I then include a chapter on tropical birds and one on other animals: the mammals, reptiles, amphibians, and arthropods. These chapters are meant to acquaint you with some of the creatures most commonly encountered in trips south of the border. Two brief chapters are included on habitats other than rain forest, one on savanna and one on coastal ecosystems. I close with a chapter dealing with conservation issues in the tropics.

One researcher in particular is deserving of special mention because he has contributed so much to the field of tropical ecology. Without the work of Daniel Janzen, this book would have been very different and far less insightful.

I am very grateful to Robert A. Askins, Nicholas V. L. Brokaw, Stephen P. Hubbell, and Leslie Johnson, each of whom critically read various parts of the manuscript and provided numerous excellent suggestions. Robert Askins also provided useful material on ancient Maya farming techniques and modern hydraulic agriculture, as well as insightful discussion on tropical diversity and its possible causes. I also wish to thank William E. Davis, Jr. for his comments on chapters 4 and 6 and Brian E. T. Cassie for his comments on chapter 2. William Gotwald supplied information on army ants, for which I am grateful. Burkhard Seubert provided information on Alexander von Humboldt.

My student workers, Mary Rockwell, Marcela Alverez, and Ivana Magovcevic, were most helpful with such tasks as loading text into the word processor and collation of material.

This book could not have been written without first-hand experience in the tropics. My peregrinations in Belize, Guatemala, Panama, Trinidad, Tobago, Puerto Rico, Peru, and Ecuador have been made possible by a variety of individuals and granting agencies. I am very grateful to Frederick J. Dodd of International Zoological Expeditions for his immense help and friendship, especially in the early days of the Belize-Guatemala trips. My thanks also to the Bowman family for their hospitality and friendship in Belize. Grants from Wheaton College, Andrew Mellon Foundation, and The Center for Field Research (Earthwatch) provided funds for many of my trips. Part of the book was written during a brief stay at Oxford University, made possible by a grant from GTE/Focus to Wheaton College. Stephanie Gallagher of the Oceanics School made it possible for me to visit Peru and Ecuador. I thank the many participants of my various Chautauqua courses for their enthusiasm and animated discussions. Ann Spearing and William Zeitler were both instrumental in providing me with the opportunity to participate in the Chautauqua program. Among my field companions, I wish to thank William E. (Ted) Davis, Wayne Petersen, Melinda Welton, and the Oceanics (Helen, Barbara, Mary Beth, Steve, Andy, Lyman, Chris, Kate, Jon, Enrique,

Charo, and Scott) for providing such memorable and downright good times. My many other traveling companions, both students and colleagues, are too numerous to mention, but all of you made it fun.

Deserving of special thanks for her contribution to this book is the artist, Andrea S. LeJeune. Her talent and hard work have made the book a visual delight.

The book was originally suggested by Mary Kennan, and I am grateful for her encouragement and support. I also thank Judith May of Princeton University Press for her support. John A. Gwynne, Jr., kindly permitted Ms. LeJeune to redraw several birds from *Birds of Panama*.

My wife, Linda, tolerated me without complaint as this book was undergoing its gestation period. She has also been a terrific field companion.

The tropics are changing rapidly, as rain forests are cleared to make room for agriculture, pasture, and other human activities. The speed with which the rain forests are being lost is a matter of deep concern to many biologists because there is so much yet to learn. I am one who hopes this pace can be slowed. I hope that when you complete your reading of this book, you'll not only understand the American tropics better but will also share my concern for and awareness of the need to preserve the uniqueness of tropical America.

A NEOTROPICAL COMPANION

CHAPTER 1

Tropical Ecosystems

NEVER does nature seem more bountiful than in the tropics. For anyone sympathetic with the pursuit of natural history, a visit to a tropical region is a must. This is a book about the New World or *Neotropics*. Alexander von Humboldt, Henry Walter Bates, Charles Darwin, Alfred Russel Wallace, Louis Agassiz, Thomas Belt, Charles Waterton, William Beebe, Frank M. Chapman, and other eminent naturalists have been profoundly influenced in their beliefs about natural history by visits to the Neotropics. Their spirits of adventure and investigation are no less fervent today. Thousands of tourists annually travel to Neotropical jungles and forests in search of birds, colorful butterflies, and other attractions of these unique ecosystems. Students and professional researchers by the dozens are unraveling perhaps the most complex Gordian knot in ecology, the multitudes of interactions among plants and animals of tropical forests. There is an urgency about the science of tropical ecology: tropical forests today are being cleared at ever-accelerating rates. Cattle pastures are replacing rain forests. The ecosystems that comprise the main subject of this book are increasingly endangered. These ecosystems deserve better. Alexander von Humboldt, one of first of the great naturalists to learn from the tropics, captured the sense of wonder one receives upon seeing rain forest for the first time:

> An enormous wood spread out at our feet that reached down to the ocean; the tree-tops, hung about with lianas, and crowned with great bushes of flowers, spread out like a great carpet, the dark green of which seemed to gleam in contrast to the light. We were all the more impressed by this sight because it was the first time that we had come across a mass of tropical vege-

tation. . . . But more beautiful still than all the wonders individually is the impression conveyed by the whole of this vigorous, luxuriant and yet light, cheering and mild nature in its entirety. I can tell that I shall be very happy here and that such impressions will often cheer me in the future. (Quoted in Meyer-Abich 1969)

Most people who have never been to equatorial regions assume the area to be continuous rain forest, much as described by the Humboldt quote. Tropical rain forest is, indeed, a principal ecosystem throughout much of the area and is the major focus of this book. However, other ecosystems also characterize the tropics (Beard 1944; Holdridge 1967; Walter 1971). Climate is generally warm and wet but is by no means uniform, and both seasonality and topography have marked effects on the characteristics of various tropical ecosystems. In this chapter I will present an overview of the tropical climate, seasonality, and major ecosystem types occurring in the Neotropics.

The Climate

Definition of the "Tropics"

Geographically, the tropics are equatorial, the area between the Tropic of Cancer and the Tropic of Capricorn, approximately a 50 degree band of latitude that, at either extreme, is subtropical rather than tropical. Tropical areas are within the trade wind belts except near the equator, an area known as the intertropical convergence or doldrums, where winds are usually quite mild. From the equator to 30 degrees north latitude, the eastern trade winds blow steadily from the northeast, a direction determined because of the constant rotation of the earth from west to east. South of the equator to 30 degrees south latitude, the eastern trades blow from the southeast, again due to the rotational motion of the planet. The changing heat pattern of air masses around the intertropical convergence causes seasonal rainfall. From early sum-

Tropical rain forest (shaded areas) in the Neotropics.

mer throughout autumn, severe wind and rain storms called hurricanes can occur.

Seasonal variations in day length are not dramatic in the tropics. At the equator, a day lasts exactly twelve hours throughout the year. North of the equator, days become a little longer in summer and shorter in winter, but this only means that summer sunset is at 6:15 or 6:20 rather that 6:00 P.M.

Temperature also fluctuates very little. Typically, the

temperature averages between 80° and 85°F with virtually no seasonal fluctuation.

El Niño

South American and, indeed, global climate is periodically and unpredictably affected by an event called *El Niño* ("The Child"). So named because it often arrives in South America around Christmas time, El Niño causes major disruptions to ecosystems. El Niño occurs when a high pressure weather system that is normally stable over the eastern Pacific Ocean breaks down, destroying the pattern of trade winds. Warm waters flow along the normally cold South American coast, global heat patterns change, and weather systems change, causing floods, droughts, fires, and other disastrous effects. In 1982, El Niño is estimated to have caused $8.65 billion worth of damage worldwide. There have been eight major El Niño events since 1945. The causal factors responsible for the periodicity of El Niños are unknown (Canby 1984; Graham and White 1988). Tropical ecosystems, already sensitive to seasonal variation (next section), are severely affected by changes caused by El Niño (see Foster 1982b, below).

The Importance of Seasonality

Rainfall varies seasonally in tropical latitudes. Because of warm air throughout the year, precipitation is in the form of rain (except atop high mountains such as the Andes, where snow occurs), but the amount of rain varies considerably from month to month. Though some rain falls each month, there is a pronounced wet and dry season. Dry season is defined as less than 10 centimeters of rainfall per month, and rainy season features between 25 centimeters and 100 centimeters of rainfall per month. The rainy season varies in time of onset, duration, and severity from one area to another in the tropics. For example, at Belém, Brazil, virtually on the equator, dry season months are normally August through November, and the wettest months are January through April. In Belize City, Belize, at 17 degrees north latitude, the rainy season begins moderately in early June but in earnest in mid-July and lasts through mid-December and sometimes into January. The dry months are normally mid-February through May.

The seasonal shift from rainy to dry seasons has direct effects on plants and animals inhabiting rain forests as well as other tropical ecosystems. One common misconception about the tropics is that seasonality can be ignored. Images of year-round sunny skies and soft trade winds are the stuff of myths. The truth is that seasonal shifts are often quite marked and many ecological patterns reflect responses to seasonal changes. During the rainy season, skies are cloudy for most of the day and heavy showers are intermittent, often becoming especially torrential during late afternoon and evening. In the dry season, skies are clear for up to ten hours during the day, and showers, though sometimes heavy, are of brief duration. These differences are not trivial to organisms. Henry Walter Bates in *The Naturalist on the River Amazons*, published in 1892, wrote of seasonal patterns as they affect forest life. Recent studies, particularly those carried out by researchers on Barro Colorado Island (BCI) in the Canal Zone of Panama have documented the drama of the changing seasons of the tropical forest (Leigh et al. 1982).

Trees flower more commonly during the dry season (Janzen 1967, 1975). With less frequent and intense showers, insects are active for longer periods, thus enhancing cross-pollination. Some tree species synchronize their flowering after downpours (Augspurger 1982), which may increase pollination efficiency by concentrating the number of pollinators (Janzen 1975).

Dry season pollination also enables more seedlings to survive because they sprout at the onset of rainy season, when there is adequate moisture available for their initial growth. Nancy C. Garwood (1982) studied 185 plant species on Barro Colorado Island and found that most seedlings emerged within the first two months of the eight-month rainy season. Forty-two percent of the plant species underwent seed dispersal during dry season and germination at the onset of rainy season. Forty percent of the species experienced seed dispersal at the beginning of rainy season and germination later in rainy season. Approximately eighteen percent of the species produced seeds that were dispersed during one rainy season, dormant during the next dry season, and germinated at the onset of the second rainy season. The species most sensitive to onset of the rainy season were "pioneer" tree species, lianas, canopy species, and both wind- and animal-dispersed species. Understory and shade-tolerant species were less sensitive.

Fruiting patterns on Barro Colorado Island were also seasonally influenced (Foster 1982a). The timing of fruiting in many species appeared to be a "compromise" between the desirability of seeds germinating at the onset of rainy season versus the advantages of flowering early in the rainy season, when insects are most abundant (see below).

Pioneer tree species often germinate at the onset of rainy season just before tree falls tend to be most common, opening gaps in the forest where these species can become established (see chapter 3). Nicholas V. L. Brokaw (1982), who studied seasonal tree fall patterns on BCI, concluded that large gaps are created approximately one per hectare every 5.3 years, a frequency sufficient to support a high population of quick-growing pioneer tree species.

Grazing rates on leaves are more than twice as high during the rainy as during the dry season (Coley 1982, 1983). New leaves are more vulnerable to insect herbivores because they lack protective tissues and chemicals (see chapter 6). Most trees grow their new leaves during early rainy season. Some trees are deciduous during the dry season, dropping their leaves entirely.

As might be expected, arthropods, many of which are

highly dependent on plants, also show seasonal changes in abundance. A study conducted among several habitats in southeastern Peru showed that forest floor arthropod biomass was most abundant during the wet season. Virtually all arthropod taxa showed clear seasonal patterns (Pearson and Derr 1986). A similar pattern occurred on Barro Colorado Island (Levings and Windsor 1982).

Rain forest birds are also sensitive to seasonal rhythms. For many years James Karr has surveyed bird communities along Pipeline Road (see chapter 2) and on Barro Colorado Island in Panama and has documented pronounced seasonal changes in the understory bird community composition (Karr 1976; Karr et al. 1982). Manakins, small birds that feed almost entirely on fruit (see page 245), have been found not to breed during seasons of fruit shortage, and, at least at one location near BCI, the manakin population is precisely adjusted to fruit availability (Worthington 1982). A study conducted on Puerto Rico by John Faaborg (1982) concluded that for birds to successfully breed during their normal season of April–July, there needed to be adequate rainfall. Charles Leck (1972), working in Panama, noted seasonal changes in distribution and abundance of nectar- and fruit-eating birds. On Grenada, Joseph M. Wunderle, Jr. (1982) showed that the bananaquit (*Coerba flaveola*), a small nectar-feeding bird (see page 231), synchronizes its breeding to coincide with the onset of the wet season.

The tamandua (*Tamandua mexicana*), a common forest anteater, shifts its diet from ants in rainy season to termites in dry season (Lubin and Montgomery 1981). Termites are juicier than ants and so afford a higher moisture content to the anteater. Termites (*Nasutitermes* sp.) are also attuned to the seasons, swarming during the onset of rainy season (Lubin 1983). The mass emergence may insure that each swarming insect has a better chance of reproduction, because it is more likely to quickly encounter another termite. Also, potential termite predators cannot possibly eat all of the swarming masses. Thus some termites survive to initiate new colonies.

Many animals such as monkeys, cats, iguanas, and various lizards abandon deciduous forests during dry season when leaves have dropped. These creatures move to riverine gallery forests, which remain in leaf.

Nicholas Smythe and two colleagues (1982) on Barro Colorado Island learned that shortage of fruits at the end of the wet season affects the ecology of two common rain forest rodents. The agouti (*Dasyprocta punctata*), a small diurnal (daytime-active) rodent, depends on relocating seeds that it has buried to sustain itself through the months of the dry season. Another rodent, the nocturnal paca (*Cuniculis paca*), survives the dry season by browsing more intensively on leaves and living off its stored fat. Both agouti and paca forage for longer periods during dry season and their populations are indirectly limited by the dry season food shortage. Because they must forage for longer periods and take greater risks to satisfy their hunger, they fall victim to predators more frequently.

A most extreme case of seasonal stress was documented, again for Barro Colorado Island, by Robin B. Foster (1982b). Two fruiting peaks normally occur annually, one in early rainy season and one in mid-rainy season. During an El Niño year (see above), however, the second peak failed to occur. Between August 1970 and February 1971 only one-third the normal amount of fruit fell, thus creating what Foster called a "famine." Not all plant species failed to produce a second fruit crop but enough did to severely affect the animal community. Researchers on BCI noted that normally wary collared peccaries (*Tayassu tajacu*), coatimundis (*Nasua narica*), agoutis, tapirs (*Tapirus bairdii*), and kinkajous (*Potos flavus*) made frequent visits to the laboratory area to get food that had been put out for them. Foster reported that the peccaries seemed emaciated and a kinkajou looked to be starving when it first appeared. Most amazing were the monkeys. To quote Foster, "The spider monkeys, which normally visit the laboratory clearing at least once every day, now launched an all-out assault on food resources inside the buildings, learning for the first time to open doors and make quick forays to the dining room table, where they sought bread and bananas, ignoring the meat, potatoes, and canned fruit cocktail, and brushing aside the startled biologists at their dinner." Foster noted that dead animals were encountered much more frequently than in previous years. "The most abundant carcasses were those of coatis, agoutis, peccaries, howler monkeys, opossums, armadillos, and porcupines; there were only occasional dead two-toed sloths, three-toed sloths, white-faced monkeys, and pacas. At times it was difficult to avoid the stench: neither the turkey vultures nor the black vultures seemed able to keep up with the abundance of carcasses." The reason why the two sloth species, the white-faced monkeys (*Cebus capucinus*), and the pacas were less affected is because they feed on foliage. It was fruit that was in short supply. Foster documented numerous dietary shifts among the starving animals.

Studies by Foster and others cited above contrast strongly with the view of the tropics expressed in the Humboldt quotation at the beginning of the chapter. The tropics may appear luxuriant at first glance when in reality imposing significant seasonal stresses upon the plant and animal inhabitants. Furthermore, the tropics do not host stable, unchanging ecosystems. Tropical ecology is very dynamic.

The Importance of Mountains

The Andes Mountains originated approximately 65 million years ago with a geological event called the Laramide Revolution, that also created the Alps, Himalayas, and Rockies. Geologists now generally agree that earth's crust consists of huge basaltic plates that continually move, often in opposition to one another. Granite continents sit atop these plates. The South American plate, containing the continent of South America, split from the African plate about 100 million years ago, creating the south Atlantic Ocean. Since then the South American plate has been moving westward. Eventually it met the eastward-moving Nazca plate, containing the southeastern Pacific Ocean. When the two plates collided in earnest, beginning about 65 million years ago, the Nazca plate began sliding under the South American plate creating the Andes Mountains. This process, called subduction, continues today and is responsible for the geological activity evident in the volcanism and earthquakes that characterize the western part of South and Central America (Dietz and Holden 1972).

The chains of mountains stretching from southernmost Patagonia north through Mexico add to the climatic and, thus, to the biotic diversity of the Neotropics. Located in western South and Central America, the geologically youthful Andes and Mexican Cordilleras host differing altitudinal ecosystems and also serve as barriers that isolate populations, thus enhancing the speciation process (see page 147).

The north-south orientation of the Andes results in coastal Peru and Chile having some of the driest deserts in the western hemisphere. I walked the desert at Paracas, Peru, and could find neither plant nor animal on the hot soil. The Andes Mountains act like a gigantic wall preventing moisture-laden air of the Amazon Basin from reaching the Peruvian and Chilean coasts. When the clouds can't rise over the mountains it rains heavily on the eastern slope of the Andes. The air passing over the mountains is depleted of its moisture; thus, dry deserts occur on the western side. This is called a *rain shadow effect*, and consequently, ecosystems are very different, though the altitude may be the same, from one side of a mountain to the other.

On a bus ride over a mountain in northern Peru between Jaén and Chiclayo I saw a rain shadow very clearly. As our driver engineered his way up the steep mountain slope, we passed through cactus and shrubby desert. Approaching the crest, however, the clear air and blue skies gave way to misty overcast. Tall columnar cactus plants appeared, laden with bromeliads. At the crest, we were in a small cloud forest of permanent fog whose ghostlike, stunted trees were adorned with all manner of orchids, bromeliads, and other *air plants* (epiphytes). As we descended the other side, the skies clouded and rain commenced. We left the tiny cloud forest, passing through rich coffee plantations and thick forests at the same altitudes where desert had occurred on the opposite side of the mountain.

Ironically, the oceanic ecosystem off the coast of Paracas is perhaps the richest in the world. Winds blow away surface water of the cold Humboldt current, creating a condition called upwelling, the rise to the surface of cold subantarctic water loaded with nutrients and oxygen. The hordes of tiny plankton supported by upwelling are food for vast numbers of sardinelike anchovetas (*Engraulis ringens*), which supported a very successful fishing industry until poor fishery management combined with effects of El Niños resulted in an anchoveta crash (Idyll 1973; Canby 1984).

Ecosystems also change, often dramatically, from the base to the top of a mountain. Working in the western United States in the late 1800s, C. Hart Merriam described what he termed *life zones*, distinct bands of vegetation wrapped around a mountain. Creosote bush and cactus desert or *Lower Sonoran* life zone is replaced by pinyon pine-juniper or *Upper Sonoran* life zone, which is followed by ponderosa pine *Transition zone*, this giving way to spruce and fir of the *Canadian* and *Hudsonian* life zones. Zonation may appear to be quite sharp, but in reality one life zone gradually changes into another. Life zones occur because altitude results in changing climatic conditions that favor different sets of species.

South American mountains also exhibit zonation patterns, noted in detail by Humboldt in the early nineteenth century. Though Merriam's life zone concept is well known, Humboldt actually preceded Merriam in describing the concept (Morrison 1976). He carefully documented how lowland rain forest gradually changes to montane rain forest, becoming cloud forest at higher altitudes. At its altitudinal extreme, cloud forest may be stunted, becoming a pygmy forest of short, gnarled, epiphyte-covered trees. Higher still on some mountains is treeless paramo, an alpine shrubland, or puna, an alpine grassland. Zonation patterns are often complex. For example, in southern Peru, near Cuzco, I ascended to about 14,000 feet elevation and found a wet acid heathland of orchids, heather, and sphagnum moss intermingled with paramo and puna (see below).

Major Tropical Ecosystems

Hylaea—The Tropical Rain Forest

"Here no one who has any feeling of the magnificent and the sublime can be disappointed; the sombre shade, scarce illuminated by a single direct ray even of the tropical sun, the enormous size and height of the trees, most of which rise like huge columns a hundred feet or more without throwing out a single branch, the strange buttresses around the base of some, the spiny or furrowed stems of others, the curious and even extraordinary creepers and climbers which wind around them, hanging in long festoons from branch to branch, sometimes curling and twisting on the ground like great serpents, then mounting to the very tops of the trees, thence throwing down roots and fibres which hang waving in the air, or twisting round each other form ropes and cables of every variety of size and often of the most perfect regularity. These, and many other novel features—the parasitic plants growing on the trunks and branches, the wonderful variety of the foliage, the strange fruits and seeds that lie rotting on the ground—taken altogether surpass description, and produce feelings in the beholder of admiration and awe. It is here, too, that the rarest birds, the most lovely insects, and the most interesting mammals and reptiles are to be found. Here lurk the jaguar and the boa-constrictor, and here amid the densest shade the bell-bird tolls his peal." So wrote Alfred Russel Wallace

Inside a rain forest. The prominent tree in the
foreground is strangler fig.

(1895), who is credited along with Charles Darwin for discovering the theory of natural selection (chapter 4).

The Neotropical rain forest was first described by Alexander von Humboldt, who called it *hylaea*, meaning "forest" in Greek (Richards 1952). The rain forest is what most of this book is about, so I will merely define it here.

A rain forest is essentially a nonseasonal forest, where rainfall is both abundant and constant. Most of the tropics consist of, however, forests where *seasonal* variation in rainfall is both typical and important. Technically, a forest with abundant but seasonal rainfall is called a *moist* forest. The National Research Council has recently offered this definition of a tropical moist (rather than rain) forest: An evergreen or partly evergreen (*some* trees may be deciduous) forest receiving not less than 100 millimeters precipitation in any month for two out of three years, frost-free, and with an annual temperature of 24°C or more (Myers 1980). Since the term *rain forest* is in common usage, in this book I will continue to refer to lush, moist tropical forests, seasonal or not, as rain forests. I've been in many, and, believe me, it rains a lot.

The "Jungle"—Disturbed Forest Areas

When rain forest is disturbed, such as by hurricane, lightning strike, isolated tree fall, or human activity, the disturbed area is opened, permitting the penetration of large amounts of light. Fast-growing plant species intolerant of shade are temporarily favored and a tangle of thin-boled trees, shrubs, and vines results. Like a huge, dense pile carpet, a mass of greenery, or "jungle," soon covers the gap created by the disturbance. To penetrate a jungle requires the skilled use of that most important of all tropical tools, the machete. Jungles are *successional*; they will eventually return to shaded forest as slower-growing species outcompete colonizing species. I will discuss jungles and ecological succession in the tropics in detail in chapter 3.

Varzea—The Riverine Flood Plain Ecosystem

Beginning at an elevation of about 17,000 feet in the Andes only 120 miles from the Pacific Ocean, the Amazon flows almost 4,000 miles to the Atlantic Ocean. Though the Nile ranks as the world's longest river, the Amazon carries by far the world's largest volume of water. It is estimated that 20% of all river water in the world passes through the 200 mile-wide mouth of the Amazon. The river itself is 7 miles wide as far as 1000 miles upriver and large ships can pass each other for over 2000 miles (Bates 1964).

The Amazon Basin, drained by the Amazon River and its gigantic tributaries, covers an area of about 2.5 million square miles, approximately 40% of South America. Approximately 1100 tributaries service the main river and some of them, like the Rio Negro, rank as major rivers. Amazon tributaries vary in color from cloudy yellow to clear black depending upon where they originate (see page 75). These tributaries are like "interstate highways" providing access to the interior. Cities such as Iquitos, Peru, are approachable only by boat or airplane. There are no roads.

The rainy season brings floods, transporting eroded soil and minerals from mountain areas and depositing them along riverbank flood plains. Where rivers flow there is continual adequate soil moisture, and, hence, evergreen forest lines the banks. Forests that border rivers are termed *gallery forests*, but in Amazonia, the term *varzea* is used for flood plain forests occurring sufficiently far from the river's edge to support a high diversity of trees. Forests off the flood plain are called *terra firme*.

Approximately 4% of the forest area in the Amazon Basin is flood plain varzea. Many of these forests receive rich sediment from the Andes during the time of flood. The flood plain may extend up to 50 miles from the river bank. During the wet season, the river may rise between 25 and 50 feet. Whole islands of vegetation are torn loose from the banks and drift downriver. Quiet pools are covered by the giant waterlily (*Victoria amazonica*), a huge six-foot-wide lily pad with upright edges.

The banks of riverine forest support an exciting diversity of animals. Capybara (*Hydrochoerus hydrochaeris*), the world's largest rodent, graze like herds of aquatic pigs along the water's edge, ever watchful to avoid falling prey to an alligatorlike speckled cayman (*Caiman sclerops*) or perhaps an anaconda (*Eunectes murinus*), the largest of the constrictor snakes (see chapter 7). Turtles, such as the matamata (*Chelys fimbriata*) or the giant arran turtle (*Podocnemis expansa*), bask on the riverbanks. The latter can weigh in excess of 100 pounds. Riverine birds, such as the turkey-sized horned screamer (*Anhima cornuta*) and prehistoric-looking hoatzin (*Opisthocomus hoatzin*), nest along the banks (see chapter 6). J. V. Remsen, Jr. and T. A. Parker III (1983) found that 15% of the *nonaquatic* Amazon bird species are directly dependent upon riverine habitats such as beaches, sandbar scrub, river edge forest, varzea, and transitional forest. Many species depend on the annual flooding cycle to preserve the particular habitats they require.

An astounding variety of fish and other creatures inhabit the waters of the Amazon. There are more than 2400 species of fish, including many of the common tetras, characins, and catfish of the aquarist (Lowe-McConnell 1987). Schools of infamous 14-inch red piranha (*Sarrasalmus nattereri*) occur most frequently in clear water. Other potentially dangerous fish include sting rays and sharks that swim upriver from the mouth of the Amazon, electric eels (*Electrophorus electris*), which can attain lengths of 6 feet and emit a jolt of 650 volts, and a miniscule catfish called the candirú (*Vandellia cirrhosa*). Slimmer than a pencil, this tiny fish has the disconcerting habit of swimming into the human urethral opening and lodging itself with sharp spines. Removal usually requires surgery. Where it is common, this fish is, understandably, the most feared animal of the Amazon. It is normally a parasite of other fish, attaching to gills.

Less dangerous but fascinating is the South American

lungfish, *Lepidosiren paradoxa.* This eellike fish with large scales and thin ribbonlike fins can gulp air in the manner of its ancestors that swam in stagnant lakes 350 million years ago. The huge arapaima (*Arapaima gigas*) is an important protein source for the Indians along the Amazon. It reaches weights of 300 pounds and lengths of up to 10 feet. A relative of the arapaima is the smaller arawana (*Osteoglossum bicirrhosum*), an elongate fish that always seems to be looking downward. It commonly shows up in the tanks of the aquarium fancier.

Aquatic mammals also live in the Amazon. Sailing from Iquitos, Peru, I saw pairs of long-snouted, fresh water dolphins, called bouto or inia (*Inia geoffrensis*), frequently swimming around our boat. Bouto are virtually the only riverine creatures not routinely hunted by native peoples, who regard them as having a high level of intelligence as well as mermaidlike qualities, thus making them worthy of superstitious reverence.

Savanna

Part of the American tropics consists of grasslands scattered with trees and shrubs, an ecosytem called a *savanna.* Savannas may be relatively wet, like the Florida everglades, or be dry and sandy. A combination of climatic and seasonal effects, occasional natural burning, and soil characteristics probably produce savannas. Human influence also can contribute to their formation. The African plains are an immense area of savanna, but savanna is considerably less extensive in the Neotropics, where large tracts of rain forest remain. Large expanses of savanna in Central America are characterized by one tree species, Caribbean pine (*Pinus caribaea*).

Because they are open areas where heat currents rise from the ground, savannas afford ideal soaring areas for hawks and vultures. Mammals such as gray fox (*Urocyon cinereoargentus*), tayra (*Eira barbara*), giant anteater (*Myrmecophaga tridactyla*), and jaguarundi (*Herpailurus yaguaroundi*) are also easier to spot in open savanna, but Neo-

Inia, or fresh-water dolphin, found in the Amazon River.

tropical savannas totally lack the large game herds that characterize their African counterparts.

South of the Amazon Basin are the *pampas*, vast grasslands that are habitat for the ostrichlike rhea (*Rhea americana*). Savanna ecology is discussed in chapter 8.

Dry Forests: Deciduous and Thornwood

Depending upon degree of dryness, which may reflect a prolonged dry season or some other climatic condition such as a rain shadow effect (see above), tropical woodlands may be partially or totally deciduous, dropping leaves periodically. Tropical broadleaf woodlands often merge with savannas, the latter usually being drier. In

Brazil, a woodland type called *cerrado* occurs between de-
cidous forest and savanna in which trees are scattered,
semideciduous, and small, typically no taller than 25 feet.
Cerrado areas experience frequent natural fires, and the
soil is typically very sandy (Walter 1971).

Thornwoods occur in semidesert areas from Mexico
through Patagonia. Dominant trees are usually *Acacia*
species and other leguminous trees, of short stature, well
spaced apart, and often interspersed with succulents such
as cacti and agave. In many areas of thornwood, large
herds of goats can be seen wandering about. Thornwood
is very common along the Pan-American Highway
throughout Peru as well as in central Mexico and many
West Indian islands.

Montane and Cloud Forests

With increasing altitude up a mountainside, temperature decreases, but precipitation often increases, supporting cool moist forests, called montane forests. These typically are lush, with an abundance of epiphytic orchids,
bromeliads, mosses, and lycopods covering branches.
Tree ferns are often common.

Higher still, the forest becomes enshrouded in semi-
permanent mist for at least part of each day, giving rise
to the term *cloud forest*. The one I explored near Cuzco,
Peru, had clear skies throughout the morning but thick
mist from afternoon through dark. For obvious reasons,
cloud forests feel quite damp and the trees are heavily
laden with epiphytes, especially mosses, lichens, and bromeliads. As altitude increases, the forest may become
quite stunted into "elfinwoods" of short twisted trees
barely 2 meters tall. Growth is slowed by a shortage of
sunlight as well as low temperatures. Perhaps because
fungi readily invade leaves, trees produce thick and waxy
leaves. Not much energy is available for trees to invest in
stems, hence stature is short (Grubb 1977).

Montane and cloud forests support many of the same
kinds of plants and animals as are found in lowland rain

forest, but they also harbor a series of rather unique species. The spectacled bear (*Tremarctos ornatus*) dwells in cloud forests throughout the northern Andes. The only species of bear found in South America, the spectacled is named for its facial pattern of white lines surrounding its eyes and cheeks. Perhaps the most spectacular bird of the Neotropics, the resplendent quetzal (*Pharomachrus mocinno*), inhabits cloud forests (see page 219). Many of the most colorful tanagers and bush-tanagers are unique to cloud forests (Isler and Isler 1987). Some cloud forests are so remote and difficult to reach that new species of birds have been discovered in them only recently. Ornithologists have described a new species of wren (Parker and O'Neill 1985), two antpittas (Schulenberg and Williams 1982; Graves et al. 1983), and an owl (O'Neill and Graves 1977), all from Peruvian cloud forests, each an unusual occurrence for such a well-known taxonomic group as birds.

Tropical Alpine Shrubland—Paramo

Paramo is a shrubland ecosystem occurring at high altitudes throughout the Neotropics. Climate is wet and cool, often with nightly frosts throughout the year. Dominant vegetation consists of large, clumped grasses called *tussock grasses* with sharp, yellowish blades. Among the tussock grasses grow shrubs, of heights up to 15 feet, resembling small trees. Leaves grow from the base of the stem, surrounding it in a pattern termed a rosette. Most distinctive of these shrubs are the *Espeletias*, perhaps the oddest members of the immense composite family (to which daisies, asters, and goldenrods belong). Espeletias have short, woody trunks densely surrounded by withered dead leaves, topped by a rosette of thick, elongate green leaves, each covered by soft hairs that help minimize evaporative water and heat loss. Espeletias, scattered like resolute sentinels among the tussock grasses on the cold, windy Andean slopes, attract many hummingbirds and bees to feed on the nectar of the yellow flowers.

Among them is the 9-inch giant hummingbird (*Patagona gigas*), the largest of the 319 hummingbird species (see page 224).

Tropical Alpine Grassland—Puna

Puna are cold alpine grasslands replacing paramo when conditions are drier. Tussock grasses also occur abundantly on puna, as well as various succulents, such as cacti and agave. The most striking of the agaves is *Puya raimondii*, with flowers on a stalk that may attain a height of 30 feet. Puya is closely related to century plants of North American deserts and, like them, flowers infrequently. When it does, its hundreds of flowers comprise a stalk that protrudes well beyond the dense basal cluster of bayonetlike, thick leaves. Puya, as well as other puna

Puya raimondi, a characteristic species of the puna.

plants, is visited by numerous hummingbird species including the purple-collared woodstar (*Myrtis fanny*), Peruvian sheartail (*Thaumastura coa*), green-tailed trainbearer (*Lesbia nuna*), and giant hummingbird.

Puna also is the favored habitat of several interesting mammals, including the vicuña (*Vicogna vicugna*), a South American member of the camel family. Dominant males group with up to ten females in herds that roam about the barren puna (Koford 1957). Another wild camel of the Andes is the guanaco (*Lama guanacoe*). Somewhat larger than the vicuña, and a similarly humpless member of the camel family, the guanaco is perhaps ancestral to the llama (*Lama guanicoe glama*) and alpaca (*L.g. pacos*). No wild members of either of these species exist, and their origins date back to domestication by pre-Columbian peoples (Morrison 1974). Llamas and alpacas are the beasts of burden for the mountain Indians, the modern Incas. The husky mountain viscacha (*Lagidium peruanum*), a member of the rodent family, is a close relative of the famed chinchilla (*Chinchilla sp.*), a species now quite local due to overtrapping.

High above the puna and paramo, fly the black and white mountain caracara (*Phalcobaenus albogularis*) and the huge Andean condor (*Vultur gryphus*). The condor is surely the most majestic of the Andean birds as it soars effortlessly on thermal currents flying from mountaintop to seacoast. Twice the size of a turkey vulture, the condor has a 9.5-foot wingspread (surpassed, but only barely, by the largest albatrosses) and can weigh as much as 15 pounds, making it a jumbo jet among birds.

The high Andean salt lakes are habitat for several flamingo species as well as other birds rarely encountered at lower elevations. The James's flamingo (*Phoenicoparrus jamesi*) was actually considered extinct until rediscovered on a lake 14,450-feet high in the Bolivian Andes in 1957 (Morrison 1974). More common are the Andean (*P. andinus*) and Chilean flamingos (*P. ruber*). All flamingos feed on brine shrimp and other small crustaceans skimmed from the water with their peculiar hatchet-shaped bills.

Andean condors soaring.

Also inhabiting high Andean lakes are the puna teal (*Anas versicolor*) and the Andean goose (*Chloephaga melanoptera*). The former is a small duck with a black upper head and white cheeks. The latter is a stocky white goose with black on the wings and tail. It has an extremely short bill.

Along fast-flowing Andean rivers lives the torrent duck (*Merganetta armata*) and white-capped dipper (*Cinclus leucocephalus*). The sleek male torrent duck has a boldly patterned white head with black lines and sharply pointed tail. The female is rich brown. Both sexes have bright red bills. Six races of torrent duck occur from the northern Andes to the extreme southern Andes. Torrent ducks brave the most rapid rivers, swimming submerged with only their heads above water. The white-capped dipper is a chunky bird suggesting a large wren in shape. Like the torrent duck, it favors clear, cold mountain rivers, submerging itself in search of aquatic insects and crustaceans.

Distant mountain lakes often harbor unusual animals, whose populations have been isolated, allowing evolution to mold them in distinct ways. Lake Titicaca, which lies between Peru and Boliva, is home for the giant coot (*Fulica gigantea*), a ducklike bird with a chickenlike bill. In Lake Atitlán in southwestern Guatemala lives the endangered Atitlán grebe (*Podilymbus gigas*), a flightless grebe similar in appearance to, but larger than, the widespread pied-billed grebe, *P. podiceps* (Land 1970).

High Andean natural history is treated in detail by Morrison (1974, 1976).

Coastal Ecosystems—Mangroves and Seagrass

Mangroves are a group of highly salt-tolerant tree species forming the dominant vegetation along tropical coastlines, lagoons, deltas, estuaries, and cayes. Tangled forests of mangroves, some with long prop roots, others with short "air roots" protruding up from the thick sandy mud, are the nesting sites of colonies of magnificent frigatebird (man-o'-war birds) (*Fregata magnificens*), boobies (*Sula* spp.), and brown pelicans (*Pelicanus occidentalis*).

The head of an adult Andean condor.

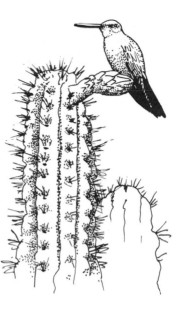

Giant hummingbird perched
atop a columnar cactus.

Flamingos.

Land hermit crabs (*Coenobita clypeatus*) leave their burrows at night to scavenge among the mangrove roots. Mangroves play a vital role in the ecology of coastal areas and contribute to the health of nearby coral reefs. Mangrove ecology is discussed in chapter 9.

Protected by the mangrove cayes, beds of seagrass cover shallow, well-lit coral sand. Like the mangroves, seagrass contributes to the health of the diverse coral reef. I will also discuss seagrass ecology in chapter 9.

The Coral Reef

The most exciting of all the coastal ecosystems is the coral reef. Approximately sixty species of coral occur in the Caribbean. Reefs of elkhorn, staghorn, finger, brain, and star corals provide habitats for myriads of colorful fish, shrimp, lobster, sea stars, brittle stars, and sea cucumbers. Coral reef ecology is discussed briefly in chapter 9.

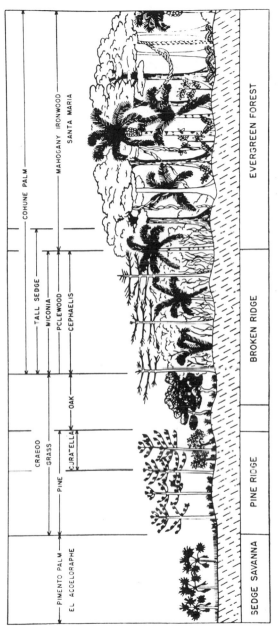

Diagram showing the range of ecosystem types present in parts
of Belize (formerly British Honduras).

CHAPTER 2

A Rain Forest: First Impressions

TALL, green, complex-looking, and surprisingly dark inside. . . . Strange bird sounds heard over a monotonous din of calling insects. . . . A screeching parrot flock dashes overhead. . . . What's up in those tree crowns, so far away? Giant trees that seem propped up by odd flaring roots. . . . No thick ground or shrub cover, actually rather easy to walk among the giant trees. . . . Occasional openings in the canopy, sunny islands surrounded by a sea of shade. . . . Dense growth in these sunny spots. . . . Vines everywhere. . . . Brilliantly colored butterflies. Lizards scampering over thick leathery leaves that seem to crackle as the reptiles scamper over them. Moist soil, rather sticky and reddish in color.

These are some typical notebook entries that might be made upon initial encounter with tropical rain forest. It doesn't matter whether you're standing in Peruvian, Brazilian, Ecuadorian, Belizean, Costa Rican, or Panamanian rain forest, it all basically looks the same. It even sounds, smells, and feels the same. All over the equatorial regions of the planet where rain forest occurs, the forest tends to have a similar physical structure and appearance. Of course, on closer inspection, numerous differences exist among rain forests both within and among various geographical areas. One does not find orangutans in Colombia nor sloths in Sumatra. Rain forests in Costa Rica are different in many ways from their counterparts in Brazil. Rain forests on poor soils differ from those on richer soils. However, the overall similarities, apparent as first impressions, are striking. Rain forests are impressive. Charles Darwin (1839) wrote of his initial impressions of tropical rain forest:

In tropical forests, when quietly walking along the shady pathways, and admiring each successive view, I wished to find language to express my ideas. Epithet after epithet was found too weak to convey to those who have not visited the intertropical regions the sensation of delight which the mind experiences.

Imagine we are standing on Pipeline Road in the Canal Zone in Panama. This gravelly, often muddy road bisects a dense rain forest and provides a convenient access to this most complex of terrestrial ecosystems. There is a well-marked trail leading off into the forest. We enter.

Structural Complexity

Once inside a rain forest, structural complexity is obvious. How dark it seems as dense canopy foliage shades the forest interior. Surprisingly, even the most colorful birds look subdued in the deep shade. Only scattered flecks of sunlight dot the forest floor. Shade has prevented a dense undergrowth from forming, and we certainly do not need our machete to move about. Plants we've seen only as potted house plants grow here in the "wild." There's a clump of *Dieffenbachia* directly ahead on the forest floor. Large philodendrons, like *Monstera deliciosa*, the "Swiss-cheese plant" with its huge, deeply lobed leaves, are growing on the trunks of many trees. The biggest trees tend to be widely-spaced, most with large, flaring buttresses. Tree boles are straight and rise a considerable distance before spreading into crowns, which, themselves, are hard to clearly discern because so much other vegetation is nesting in them. Clumps of cacti, orchids, and pineapplelike plants called bromeliads festoon the widely spreading branches. It's frustrating to try to see the delicate flowers of the orchids so high above us. Vines, some as thick as tree trunks back home, hang haphazardly, seemingly everywhere. Rounded, basketball-sized termite nests are easy to spot on the trees, and the

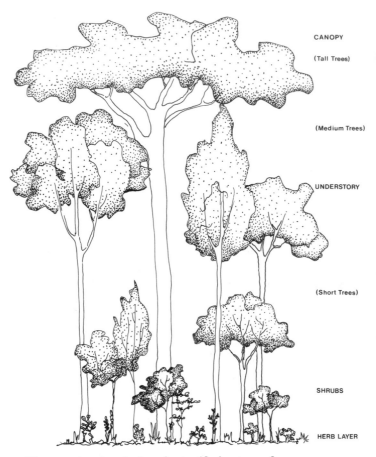

CANOPY

(Tall Trees)

(Medium Trees)

UNDERSTORY

(Short Trees)

SHRUBS

HERB LAYER

Diagram showing the loosely stratified nature of
rain forest trees. Note the widespread crowns of canopy trees
and more elliptical crowns of understory trees.

dried tunnels made by their inhabitants look like brown ski trails on the tree trunks.

North American forests are often neatly layered. There is a canopy, the height to which the tallest trees such as the oaks and maples grow, a subcanopy of understory trees such as sassafras and flowering dogwood, a shrub layer of viburnums or mountain laurel, and a herbaceous layer of ferns and wildflowers.

The tropical rain forest is not so neatly layered (Richards 1952), and up to five strata are often present (Klinge et al. 1975). The forest is sufficiently complex as to make it difficult to even discern the various strata (Hartshorn 1983a). Some trees, called emergents, erupt from the canopy to tower over the rest of the forest. Trees of varying heights abundantly comprise both understory and canopy. Most trees are monotonously green, but a few may be bursting with colorful blossoms, while others may be utterly leafless, revealing the many air plants, or epiphytes, that are attached to their branches. Shrubs and other herbaceous plants share the forest floor with numerous seedling and sapling trees, ferns, and palms. It is difficult to perceive a simple pattern in the overall structure of a rain forest. Complexity is the rule.

Typical Tropical Trees

An irony of nature in the tropics is that, though there are many different tree species, most are sufficiently similar in appearance so that one can meaningfully describe a "typical tropical tree." Trees inside a rain forest tend, at first inspection, to look alike, though an experienced observer can accurately identify many species. What follows is a general description of tropical tree characteristics. Starting from the ground up, many, if not most, trees in a rain forest have buttressed roots.

BUTTRESSES AND PROP ROOTS

A buttress is a root flaring out from the trunk to form a flangelike base. Several buttresses radiate from a given

Many tropical trees have relatively simple
branching patterns, as shown by
this individual.

tree, surrounding and seeming to support the bole, often
making cozy retreats for boa contrictors (*Boa constrictor*)
or other serpents. Buttress shape is sometimes helpful in
identifying specific trees.

The function of buttressing has been a topic of active
discussion among tropical botanists. Because buttressing
is common among trees of stream and river banks as well
as among trees lacking a deep taproot, some believe that
buttressing acts principally to support the tree (Richards
1952; Longman and Jenik 1974).

I was once told of a team of botanists in Costa Rica who

were discussing several esoteric theories for the existence of buttresses when their Indian guide offered the comment that buttresses hold up the tree. When the guide's opinion was dismissed lightly, he produced his machete and adeptly cut away each of the buttresses from a nearby small tree. He then casually pushed the tree over. Whether true or false that buttresses function principally for support, they may indeed serve other functions related to root growth patterns (page 74). Some trees lack buttresses but have stilt or prop roots that radiate from the tree's base, remaining above ground. Stilt roots are particularly common in areas such as flood plains and mangrove forests (page 346) that become periodically innundated with water.

Some tropical trees lack both buttresses and prop roots and have instead either horizontal surface roots or un-

Surface roots and buttressing characterize many tropical trees.

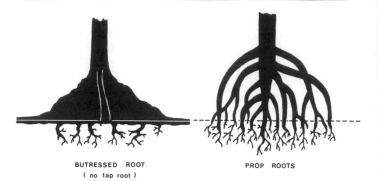

BUTRESSED ROOT
(no tap root)

PROP ROOTS

Two common patterns for root systems in tropical trees.
Prop roots are particularly common in wet areas.

derground roots. In both cases, large taproots often
occur.

TRUNKS AND CROWNS

As we look around the Panamanian forest, we notice
that many trees have tall, slender boles. The bark may be
smooth and light colored like that of the big kapok tree
(*Ceiba pentandra*) ahead of us (page 88) or dark and rough
like the giant chicle tree (*Manilkara zapota*) across the trail.
We notice some odd scars on the chicle bark, slashes made
by machetes in previous years. The chicle was the original
source of chewing gum base. Its inner pink bark, below
the dark rough outer bark, is where large amounts of
white latex, the gum base, is produced. Every two to three
years the tree is tapped by making a zigzag machete-cut
in the bark (Hartshorn 1983b). Tropical tree bark may be
thin, but in some cases it is quite thick (and the wood inside may be very hard—remember that termites abound
in the tropics). The color and taste of the underlying cam-
bium layer is sometimes a good key to identifying the tree
species (Richards 1952).

Tropical trees occasionally exceed heights of 150 feet
but more typically grow to 100 feet or slightly less. Some
emergents do top 200 feet and may occasionally approach

300 feet, though such heights are uncommon. Be aware that typical tropical trees may look taller than they really are, just as a thin person gives the appearance of being taller than a stocky person of equal height. The world's tallest trees are *not* tropical but temperate zone trees, namely, the giant sequoias and redwoods of California and eucalypts of Australia.

Many canopy trees have a spreading, flattened crown (Richards 1952). Main branches radiate out from a single point, somewhat like the spokes of an umbrella. Of course, the effect of crowding by neighboring trees can significantly modify crown shape. Single trees left standing after adjacent trees have been felled often have oddly shaped crowns. Many trees that grow both in the canopy and shaded understory have foliage that is *monolayered*, where a single dense blanket of leaves covers the tree. Trees in the understory are often lollipop-shaped and monolayered. Trees growing in forest gaps where sunlight is much more abundant (see below and next chapter) are *multilayered*, with many layers of leaves to intercept light (Horn 1971; Hartshorn 1980, 1983a).

Some trees exhibit a characteristic termed *cauliflory*, meaning the flowers and subsequent fruits abruptly grow from the trunk, rather than from the branches. Cacao, from which cocoa is produced, is a cauliflorous tree (page 103).

LEAVES

Leaves of many tropical tree species are distressingly similar in shape, making species identification difficult. The distinctive lobing patterns of North American maples and oaks are missing from most tropical trees. Instead, leaves are characteristically oval, unlobed, and often possess sharply pointed ends, called *drip tips*. Leaves tend to have smooth margins rather than "teeth." Tropical rain forest trees, mostly evergreen, produce heavy, thick, leathery, and waxy leaves that can remain on the tree for well over a year. Though many trees have simple leaves, compound leaves are by no means uncommon, especially in certain groups such as legumes. Tropical leaves rarely show obvious insect damage (see chapter 5).

FLOWERS

Many tropical trees have colorful, fragrant blossoms. Red, orange, and yellow are the most frequent colors, though lavender is also common. Some trees flower at night, producing large brilliant white flowers. Pollination is commonly facilitated by insects, birds, or bats (page 161), though wind pollination occurs in some species of canopy trees.

FRUITS AND SEEDS

Some tropical trees produce conspicuous fruits and the seeds within may be very large. For instance, the monkey pot tree (*Lecythis costaricensis*) produces thick, 8-inch diameter "cannon ball" fruits, each containing up to fifty elongate 2-inch seeds. The milk tree (*Brosimum utile*) forms succulent, sweet-tasting edible fruits, each with a single large seed inside. This tree, named for its white sap (which is drinkable), may have been planted extensively

A cauliflorous fruit, the cacao.

at places like Tikal by Mayan Indians (Flannery 1982). The famous Brazil nut comes from the forest giant, *Bertholletia excelsa*. Many tree species in the huge legume family package seeds in large pods. Among these, the stinking toe tree (*Hymenaea courbaril*) produces 5-inch oval pods with five large seeds inside. The pods drop whole to the forest floor and often fall prey to agoutis and other forest mammals. Mahogany trees (*Swietenia macrophylla* and *S. humilis*), famous for their superb quality wood, develop 6-inch oval, woody fruits, each containing about forty seeds. The seeds are wind-dispersed and would be quite vulnerable to predation were it not for the fact that they have an extremely pungent irritating taste.

Climbers, Lianas, Stranglers, and Epiphytes

As we continue our perambulations through the Panamanian rain forest we cannot help but notice the plethora of vines and epiphytes. Trees are so laden with these hitchhikers that it is often a challenge to discern the ac-

A view into the canopy showing a prominent bromeliad and many vines.

tual crown from the myriads of ancillary plants. With binoculars and practice, however, we can begin to make some sense of which is growing where and on what.

VINES

Perhaps your archetypal vision of the tropics is Tarzan and Jane swinging skillfully from tree to tree aboard a convenient vine. In reality there is little likelihood of moving through tropical forest by vine swinging unless, of course, you're a gibbon. It's true, however, that vines are a conspicuous and important component of tropical rain forests. Vines come in various forms. Some, called *lianas*, entwine elaborately as they dangle from tree crowns. Others, the climbers, attach tightly to the tree trunk and ascend. Still others, the stranglers, encircle a tree and may eventually choke it. Tropical vines occur abundantly in disturbed sunlit areas as well as in forest interiors.

A liana gets its start when a forest opening called a gap is created (page 77), and light floods in. Lianas usually begin life as shrubs rooted in the ground but eventually become vines, with woody stems as thick or thicker than the trunks of many temperate zone trees. Tendrils from the branches entwine neighboring trees, climbing upward, reaching the tree crown as both tree and liana grow. Lianas spread in the crown, and a single liana may eventually loop through several tree crowns. Lianas seem to drape limply, winding through tree crowns or hanging as loose ropes parallel to the main bole. Their stems remain rooted in the ground and are oddly shaped, often being flattened, lobed, coiled like a rope, or spiraling in a helixlike shape. The thinnest have remarkable springiness and will often support a person's weight, at least for a short time. Some liana stems are hollow, containing potable water, attainable through the use of the machete.

Francis E. Putz (1984) studied lianas in Panama and learned that a single hectare (10,000 square meters, or about 2.5 acres) hosted 1597 climbing lianas, distributed among 43% of the canopy trees. In the understory, 22%

of the upright plants were lianas, and lianas were particularly common in forest gaps. A heavy liana burden reduced the survival rate of trees, making them more likely to be toppled by winds. Fallen lianas merely grew back into other trees.

Other vines, such as the arum *Monstera deliciosa* or various philodendrons, are climbers. They begin life on the ground. Their seeds germinate and send out a tendril toward shade cast by a nearby tree. The tendril grows up the tree trunk, attaching by aerial roots, and the vine thus moves from the forest floor to become anchored on a tree.

The most notorious of the vines is the strangler fig (*Ficus crassiuscula*). This widespread tropical vine begins as a seed dropped by a bird or monkey in the tree crown among the epiphytes. Its tendrils grow toward the tree bole and grow downward around the bole, anastomosing or fusing together like a crude mesh. The strangler eventually touches ground and sends out its own root system. The host tree often dies and decomposes, leaving the strangler standing alone. The mortality of the host tree may be caused by constriction from the vine or the shading effect of the vine.

EPIPHYTES

Epiphytes are often called *air plants*. As the prefix *epi* implies, these plants live *on* other plants. They are not internally parasitic, but they do claim space on a branch where they set out roots, trap soil and dust particles, and photosynthesize as canopy residents. Rain forests, both in the temperate zone (such as the Olympic rain forests of Washington and Oregon) and in the tropics, abound with epiphytes of many different kinds. Cloud forests also host an abundance of air plants. In a lowland tropical rain forest nearly one quarter of the plant species are likely to be epiphytes (Richards 1952; Klinge et al. 1975).

Many different kinds of plants grow epiphytically. In Central and South America alone, there are estimated to be 15,500 epiphyte species (Perry 1984). Looking at a sin-

Sequence illustrating how a strangler fig replaces a palm.

gle tropical tree can reveal an amazing diversity. Lichens, liverworts, and mosses grow abundantly on both trunk and branches. Cacti, ferns, and colorful orchids line the branches. Also abundant and conspicuous on both trunk and branch alike are the bromeliads, with their sharply pointed daggerlike leaves. The density of epiphytes on a single branch is often high. I witnessed this under somewhat alarming circumstances when, following a heavy downpour, a tree limb fell from 70 feet onto my (fortunately) unoccupied tent. Though the tent was ruined, I at least enjoyed seeing the many delicate ferns and orchids growing among the dense mosses and lichens that completely covered the upper surface of the fallen branch.

Epiphytes attach firmly to a branch and survive by trapping soil particles blown to the canopy and using the captured soil as a source of nutrients such as phosphorus, calcium, and potassium. As epiphytes develop root systems, they accumulate organic matter, and thus a soil-organic litter base, termed an epiphyte mat, builds up on the tree branch. Some epiphytes have root systems containing fungi called mycorrhizae. These fungi greatly aid in the uptake of scarce minerals. Mycorrhizae are also of major importance to many trees, especially in areas with poor soil (page 73). Epiphytes efficiently take up water and thrive in areas of heavy cloud cover and mist.

Though epiphytes do not directly harm the trees on which they reside, they may indirectly affect them through competition for water and minerals. Epiphytes get "first crack" at the water dripping down through the canopy. However, Nalini Nadkarni (1981) learned that some temperate and tropical canopy trees develop aerial roots that grow into the soil mat accumulated by the epiphytes, tapping into that source of nutrients and water. Because of the epiphyte presence, the host tree benefits by obtaining nutrients from its own canopy. Donald Perry (1978) suggests that monkeys traveling regular routes through the canopy may aid in keeping branches from being overburdened by epiphytes.

Bromeliads are abundant epiphytes in virtually all Neotropical moist forests. Leaves of many species are arranged in an overlapping rosette to form a cistern that holds water and detrital material. Other species have a dense covering of hairlike trichomes on the leaves that help rapidly absorb water and minerals. The approximately 2000 New World bromeliad species are members of the pineapple family, Bromeliaceae, and, like orchids (below), not all grow as epiphytes. Those that do provide a source of moisture for many canopy dwellers. Tree frogs, mosquitos, flatworms, snails, salamanders, and even crabs complete their life cycles in the tiny aquatic habitats provided by the cuplike interiors of bromeliads. One classic study found 250 animal species occurring in bromeliads (Picado 1913, cited in Utley and Burt-Utley 1983). One group of small colorful birds called euphonias

A typical bromeliad, with flower stalk.

(page 230) use bromeliads as nest sites. Bromeliad flowers grow on a central spike and are usually bright red, attracting hummingbirds (page 224).

Orchids are a global family (Orchidaceae) abundantly represented among Neotropical epiphytes (Dressler 1981). In Costa Rica, approximately 88% of the orchid species are epiphytes, while the rest are terrestrial (Walterm 1983). Many orchids grow as vines, and many have bulbous stems (called pseudobulbs) that store water. Some have succulent leaves filled with spongy tissue and covered by a waxy cuticle to reduce evaporative water loss. All orchids depend on fungi called mycorrhizae during some phase of their life cycles. These fungi grow partly within the orchid root and facilitate uptake of water and minerals. The fungi survive by ingesting some of the orchid photosynthate; thus, the association between orchid and fungus is mutualistic: both benefit. A close look at some orchids will reveal two types of roots: those growing on the substrate and those that form a "basket," up and away from the plant. Basket roots aid in trapping leaf litter and other organic material that, when decomposed, can be used as a mineral source by the plant (Walterm 1983). Orchid flowers are among the most beautiful in the plant world. Some, like the familiar *Cattleya*, are large, while others are delicate and tiny. Binoculars help the would-be orchid observer in the rain forest. Cross-pollination is accomplished by insects, some quite specific for certain orchid species. Bees are primary pollinators of Neotropical orchids. These include long-distance fliers, like the euglossine bees that cross-pollinate orchids separated by substantial distances (Dressler 1968). Some orchid blossoms apparently mimic insects, facilitating visitation by insects intending (mistakenly) to copulate with the blossom (Darwin 1862).

High Species Richness

Looking around inside the Panamanian rain forest we cannot help but wonder how many things are looking

Cattleya orchid blossoms.

back at us. Both animal and plant life is abundant and diverse. The term *species richness* refers to how many different kinds of plants and animals inhabit a given area. In a temperate zone forest it is often possible to count the number of tree species on the fingers of both hands (though a toe or two may be needed). Even in the most diverse North American forests, those of the Appalachian coves, only about 30 species of trees occur in a hectare (10,000 square meters, or about 2.5 acres). In the tropics, however, anywhere from 40 to 100 or more species of trees can occur per hectare. Indeed, one site in the Peruvian Amazon has been found to contain approximately 300 tree species per hectare (Gentry 1988). Altogether, about 90,000 species of plants occur in the New World tropics.

British naturalist Alfred Russel Wallace, codiscoverer with Charles Darwin of the theory of natural selection (page 109), commented upon the difficulty of finding two of the same species of tree nearby each other. Wallace (1895) stated of tropical trees:

> If the traveller notices a particular species and wishes to find more like it, he may often turn his eyes in vain in every direction. Trees of varied forms, dimensions and colour are around him, but he rarely sees any one of them repeated. Time after time he goes towards a tree which looks like the one he seeks, but a closer examination proves it to be distinct.

Richness is high, and individuals within a single species often tend to be very widely scattered.

Dennis Knight (1975), working on Barro Colorado Island, very near our tract of forest on Pipeline Road, found an average of 57 tree species per 1000 square meters in mature forest and 58 species in young forest. Knight found that in the older forest, when he counted 500 trees randomly, he encountered an average of 151 species. In the younger forest, he encounted an average of 115 species in a survey of 500 trees. Stephen Hubbell and Robin Foster (1986b) have established a 50 hectare permanent study plot in old-growth forest at BCI. They surveyed approximately 238,000 woody plants with stem diameter of 1 centimeter dbh (diameter breast height) or more and found 303 species. They classified 58 species as shrubs, 60 as understory treelets, 71 species as midstory trees, and 114 as canopy and emergent trees. Both Knight's work and that of Hubbell and Foster demonstrate clearly how diverse Central American rain forests really are. Amazonian rain forests are even richer (Gentry

1988). No temperate forests begin to rival such plant species richness.

Trees are not the only diverse group. Orchids, insects, birds, and most other major groups also exhibit high species richness. A recent guide to birds of Colombia lists 1695 migrant and resident species occurring in that country (Hilty and Brown 1986). In a small area along the southern Peruvian Amazon there is a field station called Explorer's Inn where over 600 species of birds have been identified on the grounds! By comparison, barely 700 bird species occur in all of North America. More species of birds exist in the Neotropics than in the temperate zone largely because they inhabit rain forest (Tramer 1974). In Central America, bird species richness drops dramatically as soon as you leave rain forest. Tropical savannas have about the same number of bird species as similar temperate habitats (page 341).

Insect species richness can seem staggering. For the small Central American country of Costa Rica, Philip DeVries (1987) describes nearly 550 butterfly species. Edward O. Wilson (1987) reported collecting 40 genera and 135 species of ants from four forest types at Tambopata Reserve in the Peruvian Amazon. Wilson noted that 43 species of ants were found in one tree, a total approximately equal to all ant species occurring in the British Isles! Many aspects of insect species richness remain poorly known, in need of additional researchers. New species are described from virtually every collecting trip.

As we look around the Pipeline Road forest, we begin to notice how difficult it is to identify both plants and animals, in part because there are just so many different kinds.

Sights and Sounds of Animals

The rain forest, unlike the African savanna, does not provide easy views of its superabundant animal life. You really have to work at it to see animals well. Even the most gaudy birds may appear remarkably cryptic in the deep forest shade. To make matters worse, some tropical birds such as trogons and motmots sit very still for long periods and can easily be missed even when close by. Monkeys noisily scamper through the canopy, but tree crowns are so dense that we can only glimpse the simians. Iguanas remain as still as stone gargoyles as they stretch out on tree limbs high above the forest floor. The animals are there, but finding them is a different matter.

Sounds reveal the presence of some of the forest dwellers: Cicadas provide a background din reminding one of the oscillating pitch of a French ambulance siren; "Eeeee-ooooh, eeeee-ooooh, eeeee-ooooh." Parrots, hidden in the thick foliage of a fruiting fig tree, reveal themselves by an occasional harsh squeek, sounding like a door hinge in extreme need of oil. Peccaries, relatives of wild pigs, grunt back and forth to one another in low tones as they root for dinner. A trogon calls softly, "cow, cow, cow, cow." A steady, low key "threep, threep, threep" reminds us of a tree frog calling. Surprisingly, our tree frog turns out to be one of the most gaudy tropical birds, a keel-billed toucan (*Rhamphastos sulfuratus*). Nearby, a big pale-billed woodpecker (*Campephilus guatemalensis*) sounds a sharp "Rap-ta-ta-tap!"

We walk along the forest trail, careful to listen and look. A small brown animal resembling a cross between a tiny deer and a large mouse delicately prances across the trail, pausing just long enough for us to get a view of it through binoculars. We have just seen our first agouti (*Dasyprocta punctata*), a common fruit-eating rodent. We come to a stream and walk along it a short distance. Overhead, lianas hang downward. One seems very short and stiff. Binoculars reveal that it's not a vine but a tail! We've found an iguana (*Iguana iguana*). Before we have finished looking at the arboreal lizard, a bright green and rufous bird zooms by, an Amazon kingfisher (*Chloroceryle amazona*). Following slowly behind it is the large, brilliantly colored blue morpho butterfly (*Morpho peleides*), which blends perfectly against the background when resting on a tree trunk but erupts into bright blue in flight as its inner wing surfaces are revealed. The morpho only looks slow. Try to catch it with a net, but don't bet on your success.

After rejoining the trail we begin to notice the quiet. Rain forests often seem very serene, especially at midday. Were we here at night we might come upon a band of pacas (*Agouti paca*), small rodents rather similar to agoutis but with white stripes along their sides. Or we might catch sight of a family group of coati mundis (*Nasua narica*) resembling sleek, pointy-nosed raccoons. We might hear the odd treetop vocalizations of kinkajous (*Potos flavus*), another tropical member of the raccoon family. We might even encounter an ocelot (*Felis pardalis*) hunting in the cover of darkness. There are cat tracks along the streambed possibly made during last night's hunt. Finally, were we here at night there would be many kinds of bats flying about the canopy and understory. But none of these can we find, at least not easily, during the day. We look for smaller critters, searching among the leaves on the forest floor. Turning over one large leaf reveals a shiny black and orange millipede (*Nyssodesmus python*), well armored and innocuous as it slowly plods through the leaf litter.

The silence is suddenly broken by bird calls. Incredibly, birds seem everywhere when, minutes ago, there were none to be found. Soon we locate the reason for the bird flock. The trail is being crossed by a troop of *Eciton*, a large army ant. Being careful not to step where the ants are (see following section), we don't want to miss the opportunity of seeing the ant-following bird flock. Antbirds and other bird species feed on the many arthropods, the insects, spiders, and their kin flushed by the ant horde (page 259). A medium-sized soft, brown bird with a black throat and face walks methodically beside the ants, a black-faced antthrush (*Formicarius analis*). From the lower branches we spot another bird, cinnamon, with black speckles on its breast and a subtle patch of bare blue skin about the eyes, an ocellated antbird (*Phaenostictus mcleannani*). Nearby is yet another, but much tinier bird, with

Birds at an army ant swarm. From top to bottom,
the birds are a plain xenops, barred antshrike,
spotted antbird, and black-faced antthrush.

black face, black spots on a white breast, and two chestnut wing bars, a spotted antbird (*Hylophylax naevioides*). On a tree trunk we find a small woodpeckerlike bird, very rufous brown, a wedge-billed woodcreeper (*Glyphorynchus spirurus*). On the same tree is a much larger version of the same type of bird, a barred woodcreeper (*Dendrocolaptes certhia*). On a horizontal limb of a nearby small tree a large rufous motmot (*Baryphthengus martii*) sits upright, swinging its pendulumlike racket tail from side to side. Nearby a trogon flips off a branch in pursuit of a dragonfly. The trogon, seen only momentarily, had a yellow breast. Was it a violaceous (*Trogon violaceous*) or a white-tailed trogon (*T. viridis*)? We didn't get a good enough look to know. That will happen more than once. Army ants regularly attract a crowd. Before we leave the ants, we've seen a dozen bird species, and possibly more are around.

The trail has looped, bringing us out again to the Pipeline Road. Along the roadside is a clump of thin trees with huge, umbrellalike, lobed leaves. These trees seem to occur wherever an opening exists and light floods through. They are cecropias (*Cecropia* spp.), among the most abundant tree species on disturbed sites. We'll look at these in more detail later (page 82), but for now we pay little attention since, sitting idly in the midst of the largest cecropia, is a three-toed sloth (*Bradypus variegatus*). Sloths barely move, and this one is no exception. It seems just to be sitting there reacting to nothing. Slowly its left forearm moves, a parody of slow motion photography. Like the Tin Man in the Wizard of Oz before he was oiled, the sloth's muscles seem to begrudge it the ability to move.

The sloth's cecropia is flowering, the slender pendulous blossoms hanging down under the huge leaves. Soon a mixed flock of tanagers, honeycreepers, and euphonias fill its branches, gleaning both insects and nectar from the tree. Unlike the antbird flock, this group is brilliantly colored: metallic violets, greens, and reds. Beneath the cecropia grows a clump of heliconias, cousins to the bird-of-

A three-toed sloth in a cecropia tree.

paradise plant. A long-tailed hermit hummingbird (*Phaethornis superciliosus*) plunges its sickle-shaped bill into the flowers surrounded by bright orange bracts. A mild commotion along the forest edge is caused by a small troop of tamarins (*Saguinus geoffroyi*), miniature monkeys that frequent edges and areas of dense growth. Monkeys are major attractions of tropical forests, and we watch these as they investigate the various trees searching for fruits.

As we walk the road back to the car the sky begins to cloud up. The high humidity has taken its toll, and we are feeling a bit tired. One more small trail leads into the rain forest. Should we do just a little bit more exploring? It's going to rain soon, that's obvious. What the heck, we take the trail. As we approach a large buttressed tree alongside the trail, we hear an odd sound, like dry leaves buzzing. The sound is ahead of us. It's better not to go on until we locate it. Soon we find the source of the buzzing, and, in spite of the heat, it sends chills. Coiled alongside the trail, in the protection of a large buttress, is a 4-foot-long *Bothrops asper*, otherwise known as fer-de-lance or yellow-jaw. It has seen us and is vibrating its tail rattlesnake-style in the leaves. One of the most venomous of the New World snakes, this animal is to be avoided, as its bite can be lethal. It's a beautiful animal, however, and its soft browns and black diamond-shaped pattern camouflage it well against the background of shady browns. However, its distinctive triangular head and slitted catlike eyes warn us of its potential for harm. We look at the serpent from a respectful distance and carefully retreat, leaving it alone.

By the time we are back at our car pulse rates have about returned to normal. It's safe to walk in the rain forest if you know how to keep alert for possible danger. We did. Better get into the car. It just began pouring rain. My eyeglasses have steamed up. On the ride back to Panama City, we review again the dangers to keep in mind while in the tropics.

"And, Hey, Let's Be Careful Out There"

Some regard the tropics as a version of Dante's Inferno. It's hot, humid, full of scorpions, poisonous snakes, tarantulas, internal and external parasites, and disease, and, if nothing else, the marauding insects will carry you off. To these folks, the tropics is a "green hell." There are certainly potential dangers to be dealt with and, without a doubt, insects and other arthropods are there in numbers, but by and large the tropical rain forest comes closer to resembling the Garden of Eden than the Inferno. Just as you need to learn the safety rules of the road in order to enjoy driving a car, so you need to learn basic precautions when moving about in a tropical forest.

A fer-de-lance striking.

Once these become second nature to you, fear of the jungle will be a mere memory.

TRAIL PRECAUTIONS

First, don't go alone. Tropical forests are often rather remote from convenient assistance. Sounds, such as calls for help, don't travel very far in dense forest. It is a wise move to have traveling companions, a form of human insurance of help if needed. Second, know where your feet and hands are at all times. Don't carelessly step over logs (or sit on them) without looking at where your feet (or derrière) are going to land. Tall boots are really not necessary but good quality hiking shoes that will tolerate being wet will make your walking more comfortable. Many understory palms have very sharp spines radiating out horizontally from the bark, so don't just grab onto a tree without looking first. Stay on trails. I once foolishly chased a pair of chestnut woodpeckers (*Celeus elegans*) gleefully through a Trinidad rain forest. I skipped over logs, through undergrowth, and got great looks at the birds. I soon realized, however, that I had taken a chance. I was alone, well off the trail, and I had not paid too much attention to where I was going. Trinidad hosts a fair number of bushmasters (*Lachesis mutus*), very lethal snakes. I was ever so careful making my way back to the trail. Take a rain poncho, because sudden heavy downpours are routine. If you have camera and/or binoculars, take plastic bags so your equipment can be sealed and remain dry during rain showers.

POISONOUS SNAKES

Snakes are a major worry to most tropical visitors. The reality, however, is that seeing a dangerous snake is, in itself, a rather rare event. Being struck by one is considerably rarer. Robert K. Colwell (1985b) reports that 450,000 total person-hours of field work at the Organization for Tropical Studies field sites in Costa Rican rain forest occurred without one snake bite. Of course, he mentioned this figure in relation to discussing his own experience after spoiling the record by being bitten by a fer-de-lance. Our little walk in the rain forest revealed a fer-de-lance, but these snakes are not normally easy to find. Still, they are there, and one must be careful. Some poisonous snakes, like the fer-de-lance and bushmaster, are dwellers of the forest floor. Others, like the eyelash viper (*Bothrops schlegelii*), climb into low bushes and are threats to careless bushwackers. Coral snakes are potentially very lethal but rarely seen. They remain hidden under rocks and logs. Poisonous snakes are usually well camouflaged and will strike if threatened or accidently stepped upon. Keep your eyes open. Never attempt to pick up a snake if you have the slightest doubt as to its identity. Poisonous snakes are discussed further in chapter 7.

INSECT HORDES

Insects are usually not the nuisance they are reputed to be in the tropics. Anyone who has ever braved northern New England blackflies and mosquitos in spring has in all likelihood experienced worse insect problems than will be normally encountered in the tropics. True, mosquitos are present, sometimes in abundance, and sometimes large in body size. Some carry malaria, yellow fever, or other potential maladies. However, insect repellent is usually sufficient for protection. Mosquito netting or a tight tent are necessary for camping. In areas where mosquito-borne diseases occur, it is, obviously, wise to take protective innoculations in advance of the trip (malaria preventative drugs must be taken before, during, and after the trip for a period of time—see below). Mosquitos are most abundant during the rainy season and usually much less so during dry season, though sudden rains can bring forth hordes in a short time. Be prepared. There are also simulid flies related to blackflies that bite and leave little red blood spots, but these are generally not much more than a minor nuisance. A few cockroaches are nearly as big as mice and about as dangerous. In other words, "no problema."

BOTFLIES

Once, upon returning from Belize, I noticed that a small insect bite, presumably from a simulid fly, was enlarging rather than healing. Periodically the wound would feel like tiny hot needles were turning within. I was harboring a larval botfly (*Dermatobia hominis*) that had grown from an egg transported by a bloodsucking fly that had bitten me in Belize. The botfly maggot had hatched on my skin, burrowed inside, and was now using me as a source of shelter and sustenance. As it enlarges, a botfly maggot creates an obvious skin lesion, a condition called "specific myiasis" (Markell and Voge 1971). If left unattended, within forty to fifty days the larval fly will emerge to pupate. Discomfort increases as the larva grows because the insect turns, and its body is covered with sharp spines. I chose to remove the little maggot by simply covering the lesion with petroleum jelly, preventing the larva from breathing (a tiny abdominal breathing tube remains in contact with the air, though the rest of the insect is burrowed—the jelly blocked the breathing tube). I squeezed the dead larva from my arm. I was also told that botfly larvae can be removed by taping a piece of meat over the wound. The larva usually migrates from the human to the meat. Botflies are not uncommon in Latin America, and insect wounds that increase in size are suspect.

ANTS

Ants are omnipresent in the rain forest. Many are aggressive, both bite and sting, and come in large numbers. Never stand among an ant swarm crossing the trail, for they will climb on to your body and bite and/or sting you. Be warned that many ants are arboreal and may literally drop from the trees and attack. Ants also move around, especially army ants. On one occasion a traveling companion was attacked in his sleeping bag by army ants and, on another occasion, I had to remove several hundred army ants from someone's shoes after the ants selected the footware as a suitable bivouac following a night's raiding. There is one particular ant of which to beware, the giant tropical ant (*Paraponera clavata*; page 323). This inch-long black ant, sometimes called a "bullet ant," is usually a solitary forager, and it packs a mighty sting. Don't toy with it.

URTICATING CATERPILLARS

The caterpillars of many tropical butterfly and moth species are covered with sharp hairs, called urticating hairs, that cause itching, burning, and welts if they prick the skin. Do not touch any hairy caterpillars you may encounter. The fact that these hairy beasts sit on leaves and tree trunks is yet another reason for making sure you know where your hands are at all times. A friend fell down a muddy hillside in Tingo María, Peru, and felt something odd in his pocket after the fall. Reaching in he retrieved a large caterpillar covered by urticating hairs. His hand bothered him for weeks afterward.

SPIDERS, SCORPIONS, AND CENTIPEDES

Spiders, scorpions, and centipedes are very common throughout the tropics. The biggest spiders are the wolf spiders and tarantulas. Both are large hairy things the sight of which tends to scare people. Neither is a serious danger. All spiders are poisonous, but the likelihood of

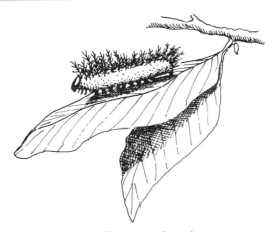

Urticating caterpillar. Note the spiny,
urticating hairs on the upper side of
the caterpillar.

being bitten is small, and, even if bitten, the likelihood of
the bite doing any real harm is remote. Wolf spiders (family Lycosidae) have long legs and smallish bodies.

Tarantulas (suborder Mygalomorpha) are thick-bodied
and hairy. They walk deliberately, almost sedately, highly
dignified for a spider, using their first pair of legs rather
like antennae, to explore their surroundings. One Amazon species, popularly called the bird-eating spider (*Eu-
rypelma soemanni*), has a 7-inch leg span. Amazon explorer
Henry Walter Bates (1892) described seeing this huge spider subduing a small bird. Though most species are large
enough to easily cover the palm of a human hand, they
bite only when highly irritated. A threatened animal will
rear up on its back legs waving its front legs at its tormen-
tor. Don't pick one up if it does this. Tarantulas have
urticating hairs on their abdomens, an additional protection against predators. Both wolf spiders and tarantulas
are nocturnal and tend to find their way into human
dwellings.

Scorpions sting rather than bite, and some, particularly
the smallest, are very toxic. Most are no more irritating

Bird-eating spider, about to eat a bird.

than the spiders, however. Though a scorpion uses its two large, lobsterlike front claws to capture prey, it defends itself (and subdues prey) by stinging. The stinger is at the tip of the last segment on its flexible abdomen. Scorpions often hide in clothing and boots, so it's best to inspect these items carefully.

Centipedes are wormlike in shape but are arthropods and move swiftly on jointed legs. They are predators and possess a pair of formidable mandibles with which the larger species can give a nasty and poisonous bite. One Central American species (*Scolopendra gigantea*) can reach lengths of 10.5 inches. Should you come upon this animal, do not pick it up.

BEES AND WASPS

Bees and wasps are abundant throughout tropical areas and, if you are allergic to the stings of such creatures, you are well advised to bring an adrenalin kit. Hospital emergency rooms often take hours to reach from field sites. Many tropical bees and wasps tend to be rather aggressive. Species of moths and other insects derive protection from mimicking bees and wasps, and certain colonial birds locate their nests near wasp nests, again to derive protection by being close to the wasps (page 232).

CHIGGERS AND TICKS

Chiggers are small mites abundant in brushy grassy areas but not as common inside rain forests. As larvae, they climb aboard a passerby and attach by inserting their mouthparts into the skin. The host tissue reacts with the chigger's saliva and itching results. Chigger bites are most common around areas where elastic from undergarments presses into the skin. Thus, the itching can be embarrassing as well as irritating. They really itch when present in numbers, and scratching can cause infection. You are advised to wear clothing that is tightly tucked in at the boots and waist to reduce chigger access. Bites should be treated with alcohol. Powdered sulfur dusted into boots and pant legs is the best preventative.

Ticks are arachnids, closely related to spiders and mites. They are ectoparasites, attaching to passing animals to feed on the host's blood. They attach by burrowing into skin with their mouthparts and must be dislodged with care so as to prevent part of the animal from remaining in the wound. Nail polish remover is helpful in dislodging ticks. Tick wounds should be treated with disinfectant to avoid infection.

INFECTION

Open wounds, even very small ones, can easily become infected in the tropics. Bacteria thrive in the hot, humid climate. Be sure to cleanse any open wound thoroughly with disinfectant. Antibiotic preparations are usually available over the counter in most Latin American countries, but care should be taken when using an antibiotic without a physician's prescription.

DOGS

Do not fraternize with dogs. Tropical dogs are malnourished, mangy, and often generously supplied with festering wounds of varying origins. They are as a general rule loaded with parasites and may easily pass eggs to humans if the human pets the dog. Tropical dogs may even be rabid, as rabies is not nearly so uncommon in the tropics as it is in the United States.

MONTEZUMA'S REVENGE

Minor diarrhea, often termed Montezuma's revenge, is caused by a "change in intestinal flora." It is to be expected, is usually temporary, and you should tolerate it with good grace and Kaopectate.

NASTY DISEASES

Though generally uncommon (for well-prepared travelers), there are certain afflictions associated with the tropics that any visitor must try to avoid.

Sanitation is often poor in tropical countries, and it is generally wise not to drink tap water or use ice made from untreated water. The "Catch-22" is that dehydration is also common in the tropics, so one must take care to ingest sufficient fluids. Bottled mineral water or commercial soft drinks are usually, but not always, free of contamination. Fruits should be peeled before being eaten. Likewise, one must be cautious about eating uncooked vegetables such as salads. Undercooked meats are to be avoided. Insect repellent, as mentioned above, should be part of every traveler's field pack. Upon return from the tropics it is wise to check any unusual symptoms promptly. Any intestinal or other odd symptoms, particularly those that may show up soon after the trip, should be checked by a physician. It is important to state clearly both when and where you were in the tropics.

Yellow fever is a potentially fatal viral disease transmitted by the mosquito *Aedes aegypti*. It was a major scourge during construction of the Panama Canal. Today yellow fever is less widespread, and inoculation is therefore needed only for certain countries.

Malaria is also a mosquito-borne disease (genus *Anopheles*) but is caused by a tiny parasitic protozoan (*Plasmodium* spp.) that infects red blood cells. After reproducing, plasmodia emerge in synchrony, rupturing red blood cells and producing symptoms of high fever and intermittent severe chills. Malaria remains relatively widespread and, unfortunately, drug-resistant strains are becoming more common. Usually, however, most malaria can be prevented by taking prophylactic treatment prior to, during, and following the trip.

Typhoid fever is a serious bacterial disease caused by *Salmonella typhosa*. It is closely associated, as are other *Salmonella*-caused illnesses, with poor sanitation. Typhoid inoculations should be taken prior to the trip.

Hepatitis is a viral disease of the liver also associated with poor sanitation. It is not uncommon, but the possibility of contracting it is reduced by taking a gamma globulin injection just prior to departure.

INTERNAL PARASITES

Parasites, such as amoebas, trypanosomes, tapeworms, flukes, and roundworms are common in areas of poor sanitation and can be very serious medical problems if not treated (Markell and Voge 1971).

A common and potentially serious amoebic infection is caused by *Entamoeba histolytica*. Called amoebic dysentery, the tiny protozoans can cause severe ulcerations of the intestines and can spread to other organ systems. Symptoms of intestinal distress such as abdominal pain and bloody stools should be checked by a physician. Amoebic dysentery is treated with drugs.

Trypanosomes are protozoan flagellates that infect the bloodstream causing serious illness. The most well known trypanosome is *Trypanosoma rhodesiense*, responsible for the infamous "sleeping sickness" of Africa. In Latin America a somewhat similar affliction, Chagas' disease, is caused by *T. cruzi*. The protozoan is vectored by many insect species, but *Panstrongylus megistus*, the reduviid bug, is considered to be the most important. Chagas' disease can be fatal, though some people recover completely and others enter remission but continue to harbor the parasite. Symptoms include high fever, heart irregularities, and digestive difficulties. Chagas' disease is treated with drugs.

Related to trypanosomes are protozoans collectively called leishmanias. The most common leishmanias in the Neotropics cause severe and rapidly spreading skin lesions, most of which begin as sores. Unusual skin conditions, sores, rashes, etc. should be inspected by a physician. Leishmanias are vectored by various species of biting sand flies. Leishmaniasis is treated with drugs, but can be difficult to cure.

Flukes are parasitic flatworms most of which occur in the Old World tropics. Some, however, are found in the Neotropics. Among these are the blood flukes. One species, *Schistosoma mansoni*, was accidently established in the Neotropics during the period of the African slave trade. *S. mansoni* causes severe intestinal distress, bloody stools, and liver degeneration. Various drugs are used to kill the parasites. Some schistosome species cause swimmer's itch. Larval schistosomes in fresh water burrow into human skin, causing local irritation.

Tapeworms are also flatworms. Most occur only in the digestive system, but one species, the hydatid worm (*Echinococcus granulosus*), forms cysts in the liver and other organs. Tapeworms are acquired by eating undercooked meats or practicing poor sanitary habits. The hydatid worm, because it invades organs such as the liver, heart, and lung, can be fatal. It is normally found in canines. It alone is reason enough not to touch stray dogs. Hydatid cysts must be removed surgically. Tapeworm infections confined to the intestine are treated with drugs.

Roundworms are common in many tropical areas. Some species infect the intestinal system, others attack the blood and other tissues. Among the most common intestinal roundworms are the hookworms, the most common of which is *Necator americanus*. Hookworms invade the body by burrowing through skin, a good reason for not going barefoot in the tropics. They eventually find their way to the intestinal system, multiply, and produce symptoms such as colic, nausea, and abdominal pain. Severe infections cause bloody stools and overall lethargy, due to blood loss. Drugs are used to eliminate the worms.

Another common roundworm, *Trichinella spiralis*, causes trichinosis, a painful and occasionally fatal muscle affliction. Ingestion of undercooked pork is the usual way in which *Trichinella* invades the human body. Drugs are used for treatment.

One roundworm group, collectively called filarial worms, causes several of the most serious diseases of the tropics, including elephantiasis and loa loa. Fortunately for the Neotropical traveler, most of these are generally confined to the Old World tropics. Some species do occur in Latin America, however, and can cause serious pathology of the subcutaneous tissues, sometimes leading to disfigurement. Larval worms can cause inflammation of the facial area, shoulders, and elsewhere, often quite painful. Drug therapy is required.

BE AWARE, BE PREPARED

Tropical disease and parasite infection can be avoided by being aware, taking all prudent precautions while in the tropics, and planning ahead. Know where your trip is taking you and what maladies might await you. Does drug-resistant malaria occur there? Does yellow fever occur there? Check with a physician familiar with tropical medicine and seek professional advice concerning appropriate innoculations and preventative medicine. See your physician promptly if, after returning, you notice any unexplained symptoms. Proper preparation and diligence will not only protect you but will give you peace of mind and confidence, adding to the enjoyment of your journey.

CHAPTER 3

How a Rain Forest Functions

THE remarkable structural complexity of tropical rain forest provides the nuts and bolts for one of the most intricate ecological machines on earth. The world's diverse rain forests capture more sunlight per unit area than any other natural ecological system. Captured solar radiation is transformed into molecules, ultimately providing structure to all organisms living within. Tropical soils, many of them delicate and mineral-poor, are nonetheless efficiently tapped for their nutrients by the hungry root systems of tropical trees. Dead plant and animal tissue quickly decays and is recycled to the living components of the ecosystem. The torrential downpours that characterize the rainy season could erode already mineral-poor soil, but the forest vegetation has ways of coping with deluges and their effects. There is much to be learned from a study of vegetation. As Alfred Russel Wallace (1895) put it,

> To the student of nature the vegetation of the tropics will ever be of surpassing interest, whether for the variety of forms and structures which it presents, for the boundless energy with which the life of plants is therein manifested, or for the help which it gives us in our search after the laws which have determined the production of such infinitely varied organisms.

Productivity

Ecologists use the term *productivity* to describe the amount of solar radiation, sunlight, converted by plants into molecules such as sugars. Plants capture red and blue wavelengths of sunlight and use the energy to split water molecules into their component atoms, hydrogen and oxygen. To do this, plants utilize the green pigment chlorophyll. Energy-enriched hydrogen from water is combined with carbon dioxide gas from the atmosphere to form high-energy sugars and related compounds. On this process, called photosynthesis, virtually all life on earth ultimately depends. Oxygen from water is given off as a byproduct. Photosynthesis, occurring over the past 3 billion years, has been responsible for changing the earth's atmosphere from one of virtually no free oxygen to its present 20% oxygen.

Of all natural, terrestrial ecosystems on earth, none accomplishes more photosynthesis than tropical rain forests. On a per unit area basis, rain forests are more than twice as productive as northern coniferous forests, half again as productive as temperate forests, and between four and five times as productive as savanna and grasslands (Whittaker 1975). As an average, a tropical rain forest annually produces about 2000 grams of dry plant material per square meter compared with 1250 grams per square meter for temperate deciduous forests (Whittaker 1975). Because of the extensive global area covered by rain forests, these ecosystems are estimated to produce 49.4 billion tons of dry organic matter annually, compared with 14.9 billion tons for temperate forests (Whittaker 1975). In the course of one year, a square meter of rain forest captures about 28,140 kilocalories of sunlight. Of this total, the plants convert 8,400 kilocalories into new growth and reproduction, using the remainder for metabolic energy.

Ecologists express leaf density as a figure called *leaf area index*, the leaf area above a square meter of forest floor. In a temperate forest such as Hubbard Brook in New Hampshire, leaf area index is nearly 6, meaning that the equivalent of 6 square meters of leaves cover one square meter of forest floor. For tropical rain forest at Barro Colorado Island in Panama, the figure is 8 (Leigh 1975). Tropical leaves also have greater biomass than temperate zone leaves. In the tropics, one hectare of dried leaves weighs about 1 ton, compared to 0.5 ton for temperate zone leaves (Leigh 1975).

Sandra Brown and Ariel E. Lugo (1982) estimated that tropical forests store 46% of the world's living terrestrial carbon and 11% of the world's soil carbon. No other ecosystem in the world stores so much carbon in living material.

One reason for the high productivity of tropical rain forests is that the growing season usually lasts considerably longer than in the temperate zone. Growth in the tropics is not interrupted by a cold winter. Temperature hardly varies, water is usually available, and, because the year is frost-free, there is no time at which all plants must become dormant, as they do in the temperate winter. The dry season does, however, slow growth and, where it is severe, many trees drop their leaves.

Productivity depends upon adequate light, moisture, and carbon dioxide, plus sufficient minerals from soil. In the first three of these essentials, tropical rain forest fares well. However, in the fourth category, sufficient minerals, rain forests are often (but not always) deprived. In many areas within the American tropics, soils are old and mineral poor, factors that could reduce productivity. However, rain forest trees have adapted well to nutrient-poor soils.

Nutrient Cycling

Because earth has no significant input of matter from space, atoms present in dead tissue must be reacquired, recycled back to living tissue. Decomposition is the process by which materials move between the living and nonliving components of an ecosystem. Bacteria and fungi convert dead tissue into simple inorganic compounds that are then available to the root systems of plants. The soil represents a temporary repository for necessary minerals such as nitrogen, calcium, magnesium, phosphorus, and potassium. Each of these minerals, and many others, are necessary for biochemical reactions in organisms, and a

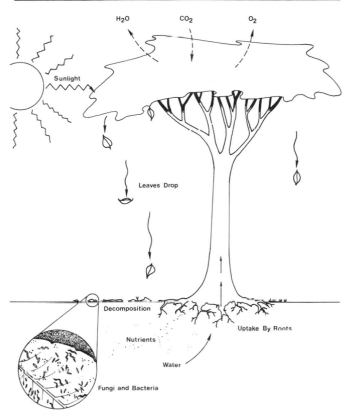

H_2O CO_2 O_2

Sunlight

Leaves Drop

Decomposition

Nutrients

Uptake By Roots

Water

Fungi and Bacteria

Diagram illustrating the basic patterns of biogeochemical cycling.

shortage of any one of them can significantly limit productivity. For example, phosphorus and nitrogen are important in the structure of nucleic acids (DNA and RNA) as well as proteins and other necessary molecules. Magnesium is part of the chlorophyll molecule, without which photosynthesis could not occur.

Consider how an atom is cycled. Suppose a dead leaf falls to the ground. Inside the leaf are billions of atoms, but we will select an atom of calcium. This calcium atom passes through several dozen fungal and bacterial species that gain energy by decomposing the leaf. Within days the calcium atom becomes part of the soil. Soon another type of fungus (called mycorrhiza, see below), growing from within a tree root, takes up the calcium. The calcium atom moves back into the tree and may well end up in another leaf. The cycle is complete and will now go around again.

Nutrient cycling is often called *biogeochemical cycling* to describe the process of chemicals moving from the bios (living) to the geos (nonliving) parts of an ecosystem. The movements of minerals in an ecosystem are strongly influenced by temperature and rainfall, the major features of climate. In the tropics, both high temperature and abundant rainfall exert profound effects on the patterns of biogeochemical cycling (Golley et al. 1969, 1975; Golley 1983).

Heat stimulates evaporation. As plants warm they evaporate water, cooling the plants and, thus, returning a great deal of water to the atmosphere in this heat-related pumping process called transpiration. Water from rainfall is taken up by plants and transpired under the stress of tropical heat. Since minerals are always taken up through roots via water, the uptake of water is essential to the uptake of minerals as well. Evaporation can be a mixed blessing. Plants can lose too much water when subjected to constant high temperature. Many tropical plants retard evaporative water loss both by closing their stomata (openings on the leaves for gas exchange) and by producing waxy leaves.

Water can wash essential minerals and other chemicals from leaves, a process called leaching. Leaching can be especially severe in areas subject to frequent heavy downpours. The protective waxy coating of tropical leaves contains lipid-soluble (but water insoluble) secondary compounds such as terpenoids that act to retard water loss and discourage both herbivores and fungi (Hubbell et al. 1983, 1984). Drip tips probably reduce leaching by speeding the water runoff. Such adaptations enable a typical tropical leaf to retain both its essential nutrients and adequate moisture.

Rainfall also leaches minerals from the soil, washing them downward into the deeper soil layers. Rainfall influences soil acidity. In the tropics, the combination of high temperatures and heavy rainfall make leaching and acid soils very common. Typical Amazon soils, termed ferralsols, are mineral-poor, high in clay, acidic, and low in phosphorus (Jordan 1982). Much water movement occurs among the atmosphere, the soil, and the organisms. Tropical plants are adapted to be very stingy about giving up minerals. Consequently, one of the major differences between the tropical and the temperate forests is that in tropical forests most of the rapidly cycling minerals are in the living plants, the biomass. Most of the phosphorus, calcium, magnesium, and potassium in an Amazon rain forest is located, not in the soil, but in the living plant tissue (Richards 1973; Jordan 1982). In the temperate zone, minerals are more equally distributed between the vegetation and soil bank.

There is normally little accumulation of dead leaves and wood on rain forest floor, making for a thin litter layer. Unlike the northern spruce-fir forests, which are endowed with a thick spongy carpet of soft fallen needles, or the temperate forests where layer after layer of fallen oak and maple leaves have accumulated, rain forest floor is only thinly covered by fallen leaves. This seems particularly odd when you keep in mind that more leaves and heavier leaves occur in rain forest. The solution to this seeming paradox is that decomposition and recycling of fallen parts occur with extreme efficiency in rain forests, much more so than in temperate forests. Studies indicate that in tropical wet forests, litter is decomposed totally in less than one year, and minerals efficiently conserved. Rain forests cycle minerals very "tightly." The resident time of an atom in the nonliving component of the ecosystem is very brief (Jordan and Herrera 1981; Jordan 1982). One study estimated that approximately 80% of the total leaf matter in an Amazon rain forest is annually returned to the soil (Klinge et al. 1975). Leaf litter does accumulate in tropical dry forests, especially during dry season (Hubbell, pers. com. 1987).

Rain Forest Soils and Mineral Cycling

It is difficult to generalize about tropical soils. In some regions, such as the eastern and central Amazon Basin, soils are very old and mineral-poor (oligotrophic), while in other regions, such as Costa Rica or the Andes, soils are eutrophic, young and mineral-rich (Jordan and Herrera 1981). Soil characteristics vary regionally because soil is the product of both geology and climate. Because these two major factors vary substantially throughout Central and South America, so do soil types. For instance, a single

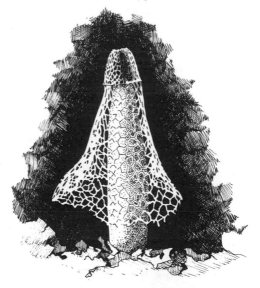

Large mushrooms, which are reproductive bodies that produce spores, are indicative of the network of fungi in tropical soils.

day's ride in southern Belize will take you from orange-red soils, the typical latisols, to clayey gray-brown soil. The gray-brown soil is largely from limestone, with which most of Belize is very well endowed.

Some tropical soils, usually red in color, are called latisols, oxisols, or ultisols. Heat and heavy moisture input cause the formation of oxides of iron and aluminum (which are not taken up by plants), giving the soil its red color. Clay content is often high, a fact quite evident as you slip and slide your way over a wet trail. Mountain roads can become more dangerous during rainy season because clay makes the roads slippery. One extreme situation that occasionally occurs with latisols results from the combined effects of intensive erosion and heat. If vegetation cover is removed and bare soil exposed to both downpours and heat, it can bake into a bricklike substance. This process, called laterization, can ruin a soil for future productivity. Attempts to farm the tropics using temperate zone methods have often been failures because of quick loss of soil fertility, sometimes due to laterization.

In parts of the Amazon Basin, soils are white and sandy. White, sandy soils are derived from the Brazilian and Guianan Shields, both very old, eroded mountain ranges. Because these soils have eroded for hundreds of millions of years, they have lost their fertility and are extremely poor in mineral content. The paradox is that lush rain forests grow on these infertile soils. I stress *on* and not in the soil because recycling occurs on the soil surface. Surface roots from the trees spread over the forest floor on which is a thin humus layer of decomposing material. A *root mat* of surface roots directly adsorbs available minerals. Carl F. Jordan and colleagues have made extensive studies of Amazon forests nutrient conservation (Jordan and Kline 1972; Jordan 1982). Using radioactive calcium and phosphorus to trace mineral uptake by vegetation, they found that 99.9% of all calcium and phosphorus was adsorbed (attached) to the root mat by mycorrhizae plus root tissue. The root mat literally grabbed and held the minerals. They also learned that roots grew very quickly.

They noted that microorganisms such as algae living within the root mat also were important in taking up available minerals. The forest floor functions like a sponge, preventing minerals from being washed from the system (Jordan and Kline 1972; Jordan and Herrera 1981; Jordan 1982).

Such a unique system may be one reason for the presence of buttresses. The buttress allows the root to spread widely at the surface, where it can reclaim minerals, without significantly reducing the anchorage of the tree. This is perhaps the tightest recycling system in nature. If the thin layer of forest humus with its mycorrhizal fungi is destroyed, this recycling system is stopped, and the fertility is lost. Removal of forest from white sandy soils can easily result in the regrowth of savanna rather than rain forest because of the destruction of the tight nutrient cycling system.

Some tropical plants have root systems that grow vertically upward, from the soil onto the stems of neighboring trees. Termed *apogeotropic roots*, these roots grow as fast as 5.6 centimeters in seventy-two hours (Sanford 1987). The advantage of growing on the stems of other trees may be that the roots can quickly and directly absorb nutrients leached from the trees, as precipitation flows down the stem. This unique system, thus far described only for some plants growing on poor quality Amazon soils, results in recycling *without* the minerals ever entering the soil!

Jordan and his colleagues also learned that canopy leaves play a direct role in taking up nutrients. Calling them nutrient scavengers, Jordan pointed out that algae and lichens that cover the leaves adsorb nutrients from rainfall, trapping the nutrients on the leaf surface. When the leaf dies and decomposes, these nutrients are taken up by the root mat and returned to the canopy trees (Jordan et al. 1979).

Some canopy plants can take up nitrogen directly from the atmosphere. Richard T. T. Forman (1975) found that certain epiphytic lichens convert gaseous nitrogen into usable form for plants in a manner similar to that of leguminous plants, which have nitrogen-fixing bacteria in their roots. Working in Colombia, Forman estimated that between 1 and 7 pounds of nitrogen per acre was supplied annually by canopy lichens. These "nitrogen-fixing" lichens provide an important and direct way for nitrogen, vital to most biochemical processes, to enter the rain forest vegetation.

White, sandy soils are usually drained by rivers that are called black waters, such as the Rio Negro. The dark, clear water is colored by tannins and phenolics, protective compounds leached from fallen leaves. I'll discuss protective compounds in detail in chapter 5, but for now I want to point out that leaves are at a great premium on such poor soils. It is not easy to find raw materials to replace a fallen leaf. Therefore, leaves on plants growing on white, sandy soils are packed with poisons that help discourage herbivory. When the old leaf finally does drop, the rainfall leaches out the tannins and phenols making the water dark, a kind of "tropical tannin-rich tea," called *blackwater*, that drains into the Rio Negro. Ecological relationships among species inhabiting blackwater forests are different in many ways from those of species in forests situated on richer soils (Janzen 1974).

In contrast to white, sandy soils, soils in Puerto Rico, many parts of Central America, and the Andes Mountains are not mineral-poor but mineral-rich. These soils are much younger, barely 60 million years old. Though exposed to high rainfall and temperature, they can be farmed efficiently and will maintain their fertility if basic soil conservation practices are used.

A dramatic example of the difference between tropical soils occurs at the confluence of the Amazon River and the Rio Negro at Manaus, Brazil. The clear, dark Rio Negro, a major tributary draining some of the white, sandy soils of the Guianan Shield, meets the muddy Amazon draining from the Andes, beginning just west of Iquitos, Peru, as a confluence of several Andes tributaries, principally the Ucayali and the Marañón. The result is a swirling maelstrom of brown and black water where the brown sediment-loaded Andes waters meet the tannin-rich clear water of an older and very different soil. The most remarkable feature is that both soil types support equally impressive rain forest.

Forest Disturbance—Ecological Succession

As you traverse the American tropics you will notice much open brushy habitat. In some places, plant cover is so dense as to be impenetrable without a guide skilled in the use of the machete. Blankets of vines wrap around thorny brush as tall, thin trees and big feathery palms push up above the tangled mass. Clumps of huge-leaved plants, named heliconias, for their sun-loving habit, compete for sunlight against scores of legumes, the most distinctive of which is the lavender-flowered mimosa, which grows in the tropics as crabgrass and dandelions grow in temperate fields. Dieffenbachia, bird-of-paradise, lantana, morning-glory, mimosa: the delights of northern horticulturists are weeds in their native land. This is the habitat we can correctly call "jungle."

Rain forests have been felled for many reasons, and evidence of forest disturbance is seemingly everywhere. During the eighteenth and nineteenth centuries, eastern North American forests were also felled so that the land could be converted to agriculture and pasture. Approximately 85% of the original New England forest was cut and in use for homestead, agriculture, or pasture at any given time during the 1800s. Now much of that forest has returned, grown back following the abandonment of agricultural land as large-scale farming moved to the midwestern states. The process whereby abandoned land returns to its original state (prior to disturbance) is called *ecological succession*. Through a succession of various species over time, an abandoned field grows back into forest. Succession is a complex process affected by many factors including chance. In the tropics, successional patterns show considerable variability (Ewel 1980).

Succession does not have to be initiated by human activity. Nature regularly disturbs ecological systems. In the tropics, heavy rainfalls, hurricanes, and high winds destroy individual old canopy trees (often bringing down neighboring trees because of lianas interconnecting their crowns) and sometimes whole forest tracts. Natural disturbances have a similar effect to plowing. A section of landscape is opened to sunlight and a whole series of plant species, each adapted to colonizing, can suddenly grow and carry out their life cycles. Accompanying the plants come many birds and other animals adapted to utilize food and cover provided by the new vegetation.

Native peoples in tropical America have long known about and exploited the tendency of the land to return to its original state following disturbance. They have adapted their agricultural practices to follow nature's pattern (page 95).

Forest Gaps

Rain forest trees periodically fall. They may be felled by humans, they may fall from windthrow, or topple when weakened by termites, epiphyte load, or old age. When a rain forest tree falls it creates a canopy opening, a *forest gap*. In gaps, light is abundant and microclimatic conditions differ from those inside closed canopy. Air and soil temperatures as well as humidity fluctuate more widely in gaps than in forest understory.

Gaps can be of many sizes and result from many causes. An emergent tree, should it fall, can take several other trees with it, creating quite a large gap. Lianas, connecting several trees together, can increase the probability of multiple tree falls. When one tree goes, its liana connection to others results in them being pulled down as well (Putz 1984). Tree falls are often correlated with seasonality. On Barro Colorado Island, tree falls peak around the middle of rainy season, when soils as well as the trees themselves are very wet and strong gusty winds blow (Brokaw 1982). Landslides can cut a entire swath of forest open. In the Stann Creek Valley in central Belize I have observed the effects of a hurricane that leveled hundreds of acres as it cut through the rain forest, a giant gap indeed.

Gaps create opportunities for sun-loving species adapted to rapid growth and reproduction. Research indicates that many species depend upon gaps (Brokaw 1985; Hubbell and Foster 1986a). Gary Hartshorn (1980) points out that nearly half of the 320 rain forest tree species in one Costa Rican forest require gaps during some stage of their life cycles. Gap-dependent species include pioneers that colonize following disturbance as well as canopy species that grow only to seedling or sapling size but no more until a gap forms. Gaps create a diverse array of microclimates, affecting light, moisture, and wind conditions (Brokaw 1985). Both horizontal and vertical heterogeniety of a forest is significantly increased by gaps.

Julie Denslow (1980) has made detailed studies of gap species and suggests that rain forest trees fall into three categories, depending upon how they respond to gaps and gap size. She recognizes (1) large-gap specialists whose seeds require high temperatures of gaps to germinate and whose seedlings are not shade tolerant, (2) small-gap specialists whose seeds germinate in shade but whose seedlings require gaps to grow to mature size, and (3) understory specialists that seem not to require gaps at all. She argues that a major component of tree species richness in the tropics is the result of each species' degree of dependence and specialization on gap size.

Denslow suggests that large-gap specialists have a significant problem, namely the relative rarity of the large gaps upon which they depend for reproduction. They partially compensate for this shortage by making large seed crops, dispersed efficiently, helping to assure that they will be there to colonize a large gap whenever and wherever it occurs. The cecropia tree (*Cecropia* spp.), which we shall discuss in detail below, is a large-gap specialist. Small-gap specialists have an easier time finding what they require because small gaps are more common.

The secret of the small-gap specialist's success is the ability to grow up only to seedling or sapling size and then stop growing and wait until a gap occurs. For many years it has been known that sapling trees of some species are capable of remaining in the understory, small but healthy, continuing their upward growth only when adequate light becomes available (Richards 1952). Understory specialists do not require gaps, growing in deep shade. As expected, their growth rates are slow compared with gap species.

Hubbell and Foster (1986a, 1986b) have censused over 600 gaps in the BCI forest in Panama. They learned not only that large gaps are less common than small gaps but also that gap size and frequency change significantly as one moves *vertically*, from forest floor to canopy. They assert that a typical gap is shaped like an inverted cone standing on its point, a pattern resulting in expansion of gap area as one moves higher in the canopy. Gap number was greatest at intermediate height (10–20 meters). Groups of species, termed *regeneration guilds*, were similar to those recognized by Denslow. Primary or mature forest tree species germinate in shade and persist as sapling juveniles until gaps become available. Early pioneer species persist for long periods as seeds in soil, germinating only when a gap occurs. Pioneers grow very quickly in strong light but are highly intolerant of shade. Late secondary species germinate in gaps and replace the early pioneers. Depending upon its gap history, a forest can consist of numerous species from each of these three regeneration guilds. Because of the abundance and distribution of gap species, Hubbell and Foster concluded that the BCI forest is changing in tree species composition.

Nicholas V. L. Brokaw (1982, 1985) studied regeneration of trees in thirty varying-sized forest gaps on Barro Colorado Island, comparing pioneer with primary species. Pioneer species produced an abundance of small seeds, usually dispersed by birds or bats, and capable of long dormancy periods. Seedlings and saplings of primary species were shade-tolerant but experienced growth spurts only when gaps formed. In contrast, Hubbell and Foster found that many of the most common species showed the same low growth rates in or out of gaps (Hubbell, pers. com. 1987). Brokaw noted that pioneer species were only found in gaps but primary species were distributed (as seedlings and saplings) throughout the forest understory. Brokaw noted that stem densities increased rapidly after gap formation, indicating an intense colonization by pioneers and growth by primary species already present.

In another study, Brokaw focused only on three pioneer species and learned that the three make up a continuum of what he called *regeneration behavior* (Brokaw 1987). One species, *Trema micrantha*, both colonized and grew very rapidly, growing up to 7 meters/year. This species only colonized during the first year of the gap. After that, it could not successfully invade, presumably due to competition with other individuals. The second species, *Cecropia insignis*, invaded mostly during the first year of the gap, but a few managed to survive that entered during the second and third year. This species grew more slowly (4.9 meters/year) than *Trema*. The third colonist was *Miconia argentea*, which grew the slowest of the three (2.5 meters/year) but was still successfully invading the gap up to seven years following gap formation. Brokaw's study reveals how the three species utilize different growth patterns to successfully reproduce within gaps.

The Jungle—Succession in the Tropics

The dictionary definition of jungle is "land overgrown with tangled vegetation, especially in the tropics" (*Oxford American Dictionary*, 1980). This definition, though descriptively accurate, does not say what a jungle is ecologically. Jungles represent large areas where rain forest has been opened, usually by cutting, an event that initiates an ecological succession. Bare land is quickly colonized by herbaceous vegetation. Seeds dormant in the soil suddenly germinate, and new seeds are brought in by wind and animals. Soon vines, shrubs, and quick-growing palms and trees are all competing for a place in the sun. The effect of this intensive competition for light and soil nutrients is the "tangled vegetation" of the definition above. Jungles resemble thick, green quilts dropped on the landscape. The density of growth among the plants is such that a machete is usually required to pass through.

Just as in rain forests, high species richness is true of successional areas, and species composition is highly variable from site to site (Bazzaz and Pickett 1980). However, it is nonetheless possible to provide a basic description and point out some of the most conspicuous and common plants seen during tropical succession (Ewel 1983). This description applies to small scale successions and successions on rich soil. On poor soil, elimination of rain forest can sometimes result in conversion of the ecosystem to savanna rather than forest (page 337).

Herbaceous weeds, grasses, and sedges of many species are first to colonize bare soil or an abandoned agricultural field. Soon these are joined by shrubs, vines, and woody vegetation. Large epiphytes are almost entirely absent from early successional areas. Plant biomass usually increases rapidly as plants compete against one another. In one Panamanian study, biomass increased from 15.3 to 57.6 dry tons per hectare from year 2 to year 6 (Bazzaz and Pickett 1980). This rapid growth reduces soil erosion as vegetation blankets and secures the soil. Recent studies in Veracruz, Mexico, by Guadalupe Williams-Linera (1983) have shown that young (10 months and 7 years old) successional areas take up nutrients as efficiently as mature rain forest. Young successional fields actually have more nutrients per unit biomass than closed canopy forests. Remarkably, the leaf area index may reach that of a closed canopy forest within 6–10 years, although the vegetation is still relatively low growing, and the species composition at that time is not at all similar to what it will be as the site returns to forest. By the time the site is about 15 years from the onset of succession, the ground conditions can be very similar to closed canopy forest, though, again, the species composition is not the same. Studies by Ewel et al. (1982) demonstrated that in only 11 months from burning, study plots underwent a succession such that vegetation attained a height of 5 meters and consisted of dense mixture of vines, shrubs, large herbs, and small trees.

Dennis Knight (1975), in his study on Barro Colorado Island in Panama, found that plant species diversity of successional areas increased rapidly during the first 15 years of succession. Diversity continued to increase, though less rapidly until 65 years. Following that, diversity still increased, though quite slowly. Knight concluded that, after 130 years of succession, the forest was still changing. Hubbell and Foster (1986a, 1986b, 1986c) note that forest at BCI is actually older, between 500–600 years old, and that it is not yet in equilibrium. They conclude that though initial succession is rapid, factors such as chance and biological uncertainty from interactions among competing tree species act to prevent establishment of an equilibrium (Hubbell and Foster 1986c). This means that BCI remains in a dynamic state, continuing to change.

During early succession, many plant species are capable of self-fertilization. These plants tend to be small in stature and produce many-seeded fruits. Such characteristics are adaptive to colonizing species. In later succession, most plants require cross-pollination, and their fruits tend to be larger, with fewer seeds. These plants, often called *equilibrium species*, are adapted to persist in the closed canopy (Opler et al. 1980).

HELICONIA

Among the most conspicuous tenants of successional areas are the *Heliconias* (*Heliconia* spp., family Helicaniaceae), recognized by their huge elongate paddle-shaped leaves and their distinctive colorful red, orange, or yellow bracts, reminiscent of lobster claws, which surround the flowers (hence the common name, "lobster-clawed" *Heliconia*). Though some *Heliconias* grow well in shade, most

Heliconia flowers,
inside conspicuous
bracts.

live up to their namesake, "sun-loving," growing in open
fields, along roadsides, forest edges, and stream banks.
The big elongate leaves are similar to those of banana
trees. Conspicuous bracts surrounding the smaller flowers attract hummingbird pollinators, especially a group
called the hermits (page 226) which have sickle-shaped
bills permitting them to dip deeply into the twenty yellow-
greenish flowers within the bracts (Stiles 1975). When several species of *Heliconia* occur together, they tend to
flower at different times, perhaps an evolutionary response to competition among *Heliconia*s for pollinators,
the hermits (Stiles 1975, 1977).

Heliconias produce green fruits, which ripen and become blue-black in approximately three months. Each
fruit contains three large, hard seeds. Birds are attracted
to heliconia fruits and are important in the plant's seed
dispersal. At Finca La Selva in Costa Rica, F. G. Stiles
(1983) reports that twenty-eight species of birds have
been observed taking the fruits of one heliconia species.
The birds digest the pulp but regurgitate the seed whole.

Heliconia seeds have a six to seven month dormancy period prior to germination, which assures that the seeds will germinate at the onset of rainy season.

PIPER

Another unmistakable jungle plant is *Piper* (*Piper* spp., family Piperaceae). Common in both successional areas and forest understory, about 500 species of *Piper* occur in the American tropics (Fleming 1983). Most are shrubs, but some species grow as herbs, and some are small trees. The Spanish name, *Candela* or *Candellillos*, means "candle" and refers to the plant's distinctive flowers, which are densely packed on an erect stalk. When immature, the flower stalk droops, but it becomes stiffened when the flowers are ripe for pollination. *Piper* flowers are pollinated by many species of bees, beetles, and fruit flies. The fruits form on the spike and are eaten, and the seeds are dispersed by one group of bats in the genus *Carollia*, called "piperphiles." Several species of *Piper* may occur on a given site, but evidence suggests that they do not all flower at the same time, thus, like *Heliconia*s, competition among them for pollinators is reduced as well as the probability of accidental hybridization (Fleming 1985a, 1985b).

Piper, showing erect flowers.

MIMOSAS AND OTHER LEGUMES

Along roadsides and in wet pastures and fields throughout the Neotropics grow mimosas, spreading, spindly shrubs and trees. Mimosas are legumes (family Leguminosae), perhaps the most diverse family of tropical plants. Virtually all terrestrial habitats in the tropics are abundantly populated by legumes including not only mimosas but acacias (*Acacia* spp.), beans, peas, and trees such as *Samanea saman, Calliandra surinamensis*, and *Caesalpinia brasiliensis* (which gave Brazil its name). The colorful flamboyant tree *Delonix regia*, a native of Madagascar, has been widely introduced in the Neotropics. Amazonian rain forests typically contain more legume species than any other plant family (Klinge et al. 1975). Legumes have compound leaves and produce seeds contained in dry pods. Many legumes have spines and some, like the sensitive plant *Mimosa pudica*, have leaves that close when touched.

Mimosa pigra, an abundant species, has round pink flowers and is unusual because it flowers early in the rainy season. The flowers, which are pollinated by bees, become 8–15 centimeter-long flattened pods that are covered by stiff hairs. Stems and leaf stalks (petioles) are spiny and are not browsed by horses or cattle. Experiments with captive native mammals such as peccaries, deer, and tapir show that these creatures refuse to eat *Mimosa* stems on the basis of odor alone (Janzen 1983b). Given its apparent unpalatability, it is easy to see why *Mimosa pigra* succeeds in open areas. Janzen (1983b) reports that seeds are spread by road construction equipment, accounting for the abundance of this species along roadsides.

CECROPIAS

One of the most conspicuous genera of trees in the Neotropics is the cecropia (*Cecropia* spp., family Moraceae). Several species are abundantly represented, each occurring in areas of large light gaps or secondary growth. Pioneer colonizing species, cecropias are well adapted to grow quickly when light becomes abundant. Studies by A.M.A. Holthuijzen and J.H.A. Boerboom (1982) in Surinam have revealed that seeds remain viable in the soil for at least two years, ready to germinate when a gap is created. Cecropia seeds are anything but rare. An average of seventy-three per square meter were present in Surinam. Because there are so many viable seeds present, cecropias sometimes completely cover a newly abandoned field or open area. They line roadsides and are abundant along forest edges and stream banks.

Cecropias are easy to recognize. They are thin-boled trees with bamboolike rings surrounding a gray trunk. Their leaves are large and deeply lobed, resembling a parasol. Leaves are whitish underneath and frequently insect-damaged. Dried, shriveled cecropia leaves that have dropped from the trees are a common roadside feature in the tropics. Some cecropias have stilted roots, but they do not form buttresses.

Cecropias are effective colonizers. In addition to having many seeds "lying in wait" in the soil, cecropias grow quickly, up to 8 feet in a year. Nickolas V. L. Brokaw recorded 4.9 meters of height growth in one year for a single cecropia. They are generally short-lived, old ones surviving about thirty years, although Hubbell reports that once established in the canopy, *Cecropia insignis* can persist nearly as long as most shade tolerant species (Hubbell, pers. com. 1987). Cecropias are usually small, rarely exceeding 60 feet in height (though Hubbell [pers. com. 1987] has measured emergent cecropias 40 meters tall) . They are intolerate of shade, their success hinging on their ability to quickly grow above the myriads of vines and herbs competing with them for space. To this end, cecropias, like many pioneer tree species, have a very simple branching pattern (Bazzaz and Pickett 1980) and leaves that hang loosely downward. Vines attempting to grow over a developing cecropia can easily be blown off by wind, though I have seen many small cecropias that were vine-covered (see below). Cecropias have hollow stems that are easy to sever with a machete. I've watched

Mayan boys effortlessly chop down 10- and 15-foot-tall cecropias. Hollow stems may be an adaptation for rapid growth in response to competition for light, as it permits the tree to devote energy to growing tall rather than to the production of wood.

Cecropias have separate male and female trees and are well adapted for mass reproductive efforts. A single female tree can produce over 900,000 seeds every time it fruits, and it can fruit often! Flowers hang in fingerlike catkins, with each flower base holding four long, whitish catkins. Research conducted by Alejandro Estrada (1984) and his colleagues in Mexico showed that forty-eight animal species, including leaf-cutting ants, iguanas, birds, and mammals, made direct use of *Cecropia obtusifolia*. A total of thirty-three species of birds from ten families, including some North American migrants, fed on flowers or fruit (page 171). I have frequently stopped by a blooming cecropia to enjoy the bird show. Such trees serve the same function for tropical birds as fast-food restaurants do for people. Mammals from bats to monkeys eat the fruit, and sloths gorge (in slow motion) on the leaves.

Estrada and his co-researchers aptly described cecropias: "Apparently, *C. obtusifolia* has traded long life, heavy investment in a few seeds, and the resulting high quality of seed dispersal, for a short life, high fecundity, a large investment in the production of thousands of seeds, extended seed dormancy, and the ability to attract a very diverse dispersal coterie that maximizes the number of seeds capable of colonizing a very specific habitat. *Cecropia* seed's ability to 'wait' for the right conditions is probably an adaptation to the rare and random occurrence, in the forest, of gaps of suitable size and light conditions sufficient to trigger germination and facilitate rapid growth." Cecropias have obviously profited from human activities, as cutting the forest provides exactly the conditions it requires.

One final note on cecropias: Beware of the ants, especially if you cut the tree down. Biting ants (*Azteca* spp.) live inside the stem. These ants feed on nectar produced at the leaf axils of the cecropia, on structures called *extra-floral nectaries*. I will describe these on page 163, but for now note that these ants sometimes protect the cecropia in a unique way. Many cecropias are free of vines or epiphytes once they've reached fair size, which is good for them since such hangers-on could significantly shade the cecropia. Janzen (1969a) observed that *Azteca* ants clip vines attempting to entwine cecropias. The plant "rewards" the ants by providing both room and board, and evolutionary mutualism (page 165). However, Andrade and Carauta (1982) noted that some cecropias hosting abundant ants are, indeed, vine covered, and that the ants seem to "patrol" only the stem and leaf nodes, not the main leaf surfaces.

THE KAPOK OR CEIBA TREE

Perhaps the most majestic Neotropical tree is the ceiba or kapok tree (*Ceiba pentandra*, family Bombacaceae), the sacred tree of the Mayan peoples. Mayans believe that souls ascend to heaven by rising up a mythical ceiba whose branches are heaven itself. Ceibas are left standing when surrounding forest is felled. Much of the look of today's tropics is a cattle pasture watched over by a lone kapok.

The ceiba is a superb-looking tree. From its buttressed roots rises a smooth gray trunk often ascending 50 meters before spreading into a wide flattened crown. Leaves are compound with five to eight leaflets dangling like fingers from a long stalk. The major branches radiate out horizontally and are usually covered with epiphytes. Many lianas typically hang from the tree.

Kapoks probably originated in the American tropics but dispersed naturally to West Africa (Baker 1983). They have been planted in Southeast Asia as well, so they are today distributed throughout the world's tropics.

Ceibas require high light intensity to grow and are most common along forest edges, river banks, and disturbed areas. Like most successional trees, they exhibit rapid growth, up to 10 feet annually. During the dry season

Ceiba pentandra, the silk cotton or kapok tree, after leaf drop. Note the abundance of epiphytes on the main branches. Single trees, such as this, are common sights as forests are increasingly cleared throughout the tropics.

they are deciduous, dropping their leaves. When leafless, masses of epiphytes stand out dramatically, silhouetted against the sky.

Leaf drop precedes flowering, and thus the flowers are well exposed to bats, their pollinators. The five-petaled flowers are white or pink and open during early evening. Their high visibility and sour odor probably combine to attract the flying mammals. Cross-pollination is facilitated since only a few flowers open each night, thus requiring two to three weeks for the entire tree to complete its flowering. Flowers close in the morning, but many insects, hummingbirds, and mammals visit the remnant flowers and seek nectar (Toledo 1977). A single ceiba may only flower every five to ten years, but each tree is capable of producing 500–4000 fruits each with approximately 200 or more seeds. A single tree can therefore produce about 800,000 seeds during one year of flowering (Baker 1983). Seeds are contained in oval fruits, which open on the tree. Each seed is surrounded by silky fibers called kapok (hence the name "kapok tree" and also "silk-cotton tree"). These fibers aid in wind-dispersing the seeds. Since the tree lacks leaves when it flowers, wind can more efficiently blow the seeds away from the parent. Seeds can remain dormant for a substantial period, germinating when exposed to high light. Large gaps are ideal for ceiba.

Kapok fibers are commercially valuable as stuffing for mattresses, upholstery, and life preservers. The tree's wood is soft, not ideal as lumber, though in some areas it is used for coffins, carvings, and canoes (Baker 1983).

Kapok leaves are extensively parasitized and grazed by insects. Leaf drop may serve not only to advertise the flowers and aid in wind-dispersing the seeds but may also help periodically rid the tree of its insect burden.

A Resilient Rain Forest—A Lesson from Tikal

Tikal, a great city of the classic period of Mayan civilization, provides an example of how rain forest can return after people abandon an area. Located on the flat Petén region of western Guatemala, Tikal was founded around 600 B.C. and flourished from about A.D. 200 until it was mysteriously abandoned around the year 900. At its peak it served as a major trade center with corn, beans, squash, chile peppers, tomatoes, pumpkins, gourds, papaya, and avocado being brought from small farms to be sold in the markets of the city. A population of 50,000 Mayas lived in Tikal, which spread over an area of 123 square kilometers, protected by earthworks and moats. The city was surrounded by densely populated suburbs probably sup-

Temple II in the Great Plaza at Tikal, Guatemala.

ported by sophisticated agriculture (LaFay 1975; Flannery 1982; Hammond 1982). The majestic pyramidlike temples, excavated relatively recently, serve as silent memorials to a long-past civilization that developed a calendar equally accurate as today's, a complex writing system that still has not been entirely deciphered, and a mathematical sophistication that included the concept of zero. Few experiences rival the sight of the Great Plaza and Temple I, the Temple of the Giant Jaguar, in the cool early morning mist. For a time, Tikal was the Paris, the London, the New York of Mesoamerica.

Today Tikal is rain forest. Its once crowded plazas, thoroughfares, and temples are overgrown by epiphyte-laden milk trees (*Brosimum alicastrum*), figs, palms, mahoganys, and chicle trees, to name but a few. From atop the sacred temples, spider and howler monkeys can be observed cavorting in the treetops. Agoutis and coatimundis shuffle through the picnic grounds, amusing tourists. Toucans and parrots pick fruit from trees growing along the central avenues leading to and from the city. Bird-watchers search the tall comb of Temple II, trying to spot nesting orange-breasted falcons (*Falco deiroleucus*). My point is that this once great metropolis of 50,000 Mayans, stretching many square miles, has been reclaimed by rain forest. Tikal was one of the largest forest gaps in the history of the American tropics, but, given hundreds of years, the gap closed.

Though many areas of rain forest (i.e., those on white, sandy soils) are very fragile, Tikal demonstrates that the process of ecological succession can restore rain forest, even after profound alteration. All of Tikal is "second growth" forest, and in some respects it is quite different from what it probably was before Mayans converted it to city and farmland. Nonetheless, it is now diverse, impressive forest. The occurrence of disturbed areas and gaps of various sizes has probably always been true of rain forests, and many species have specialized to utilize these openings. The high diversity of rain forests may be caused in large part by frequent disturbances of varying magnitudes that make it possible for a range of differently adapted species to move in and utilize the temporary conditions created by the disturbance. I will say more about this in the next chapter.

LIVING OFF THE LAND IN THE TROPICS

The native peoples of tropical America have utilized jungles and rain forests for thousands of years, well before Columbus and European culture arrived. All food and fiber are derived from the ecosystem. The culture of the people is deeply interwoven with their knowledge of the land. Native tropical Americans have learned how to create and farm their own forest gaps without the soil permanently losing fertility. They have learned which crops supply carbohydrate and which supply protein. They know how to time their plantings to make the best use of the rainy season. They know how to hunt for a variety of game and are experts at fishing. They can quickly construct a dwelling, knowing exactly which palms make the best thatch and which trees contain the best wood. Different peoples use the land in slightly different ways. Tribes such as the Jivaro and Mundurucú, which live in interior Amazonia, are not as influenced by modern methods and materials as the Central American Mayans. Even within the Amazon Basin there is a fundamental difference between peoples living on the *terre firme*, or upland areas, and those of the *varzea*, or floodplain areas. Still, it is quite possible to describe in a basic way how one goes about the business of living off the land in the tropics.

Hunting and Gathering

Native rain forest peoples have a pragmatic understanding of the land. They not only know which species of trees (named, of course, in their own language) occur in a given forest tract, but they know which are toxic and could, thus, supply arrow poison or be harmful if eaten.

They know the habits of animals, especially the peccaries, tapirs, agoutis, pacas, and various birds that make up an important protein source in their diets. In the Amazon Basin, for instance, people relish the fruit of the emergent tree *Bertholletia excelsa*, because inside each fruit are anywhere from a few to twenty Brazil nuts, rich in protein and oil. In Belize, I have continually been amazed by the ability of Mayan men to spot iguanas hidden in the foliage high above the stream bank. Sixty-year-old Ignacio, a skilled Mayan iguana hunter, adeptly climbs up the tree to shake the branch to dislodge the surprised lizard. The iguana leaps from the tree branch to the presumed safety of the stream, only to have another Mayan man dive in and retrieve it. In many areas, large game is brought down by applying poison, sometimes from the colorful frog *Dendrobates* (page 117), to the tips of arrows. Amerindians need to be and are highly skilled at handling these potentially lethal amphibians.

Hunting and gathering in its most pristine form includes no element of farming. Most hunter-gatherer populations are quite small, because there is very little actual energy directly available to the people. Most trees, epiphytes, vines, and animals are not eaten. Forest is not cut. The people blend in among the other inhabitants of the rain forest and take only what they can find and catch. The average size of a hunter-gatherer population is only about one person per square mile.

Hunter-gatherers are usually nomadic, living in a small temporary village and moving on when they have exhausted the game from a given locality. Hunting is done by blowgun (for monkeys and birds) and spear (for hoofed animals), though shotguns are now common. Some Neotropical hunter-gatherer tribes are highly territorial while resident in a given forest tract. In Brazil, Mundurucú headhunters in the recent past made no distinction between people of different tribes and animals such as peccaries and tapirs—all were hunted. Tribal warfare was probably a response to the need to protect areas of forest for the exclusive use of a single tribe, thus increasing the forest's yield (Wilson 1978). There are few remaining hunter-gatherers in the Amazon Basin. Most people use agriculture of some sort to supplement their diets.

A prevalent belief about hunter-gatherer nomadism is that tribes must periodically move because their hunting efforts deplete local game populations. William T. Vickers (1988), who studied an Indian population in tropical rain forest of northeastern Ecuador, documented the dependency of the people on local game. Three animals, collared peccary, white-lipped peccary, and tapir (chapter 7) supplied approximately 90% of the tribe's meat. Other animals, including large birds, were hunted as well. However, Vickers concluded that such factors as depletion of cedar trees (*Cedrela odorata*) for canoes and palms for thatch were more important in forcing the settlement to move. In the ten years that the settlement remained in a given area, game was not significantly depleted by the hunters.

Slash and Burn Agriculture

Agriculture, which began about 7000 B.C., involves the simplification of nature's food webs and the rechanneling of energy to humans. Instead of the sun's captured energy being dissipated among many herbivores and carnivores with just a meager amount going to humans, it is refocused on but a few crop species that humans have planted for their own use. Consequently, much more energy is available to the humans, so the human population can be ten to a hundred times larger than is the case with a hunter-gatherer lifestyle. There is, however, a cost to agriculture. In order to simplify food webs, people have to perform labor. It takes work to farm because, not only must the habitat be radically changed, but the needs of the crop plants must be met as well. Farmers need to protect their investments, weeding to remove competing species, defending against herbivorous insects and rodents, and somehow insuring that adequate nutrients from the soil will remain available. The work, done by nature for free in hunter-gatherer societies, comes with a big price tag in agricultural societies. The dividends, however, are also much larger. Agriculture basically represents a disturbance, a gap, created in the ecosystem. The farmer works to prevent the normal successional processes from closing that gap so that crops for humans can be grown. Just as successional areas are temporally unstable, changing into forest with time, so agricultural systems are intrinsically unstable systems. The farmer's labor provides the stability against nature's tendency to diversify the system. I once heard it said that if an interstellar intelligence from some distant point in our galaxy should visit earth it might be tempted to conclude that the dominant species were corn, wheat, and rice, which had combined to enslave a group of odd bipedal creatures who tend to their every need. Such is agriculture.

Tropical peoples face a challenge in attempting to farm rain forest. Most of the minerals and nutrients are not in the soil but in the biomass: the trees, lianas, and epiphytes. The problem is that to clear an area for farming, it is obviously necessary to remove the mass of vegetation. But to do so seems to doom the farming effort because the mineral-poor soil will not sustain very much in the way of crops. The way out of this dilemma is to practice what has come to be termed *swidden* or *slash and burn* agriculture.

A small plot of land is chosen by a farmer who then uses his machete and axe to cut down all of the vegetation. Trees too large to be cut are girdled, killing them. The tangled pile of vegetation is then set on fire rather than removed. Fire eliminates the leaves and wood while at the same time releasing the nutrients and minerals contained within. The soil surface is fertilized by the ash from the biomass, and the ash tends to be alkaline, raising the pH, making the soil less acidic. The farmer can then plant crops, and the soil will be fertile, but only for a few years. Rainfall will still act to erode the now exposed soil and leach minerals. The crops themselves are removed, of course, and with them go more of the minerals. The result is that fertility and, thus, yield decline steadily. In one Amazon region studied, the yield for a single village was 18 tons/hectare the first year, 13 tons/hectare the second year, and only 10 tons/hectare the third year, a reduction from the first year of 45% (Ayensu 1980). Within a few years the farmer will need to abandon his plot and allow natural succession to occur, closing the gap. The typical time sequence for swidden agriculture is to farm the plot for seven to ten years (often less) and then abandon it for at least twenty years. Ideally, the farmer will not reuse the area just abandoned for nearly a hundred years or so, permitting total recovery of the system. Slash and burn agriculture requires the rotation of sites and often results in a nomadic population, who must move around in the rain forest to find suitable plots to farm.

An experimental study in Costa Rica conducted by John Ewel and colleagues (1981) demonstrated that slash and burn does not, in the short run, degrade the soil. Ewel and colleagues cut, mulched, and burned a site that contained patches of eight- to nine-year-old forest and seventy-year-old forest. Before the burn there was approximately 8000 seeds per square meter of soil representing sixty-seven species. After the burn the figure dropped to 3000 seeds per square meter, representing thirty-seven species. Mycorrhizal fungi survived the burn and large quantities of nutrients were released to the soil following burning. The remaining seeds sprouted, and vegetation regrew vigorously on the site.

LIFE IN BLUE CREEK, A KEKCHI VILLAGE

Not all slash and burn farming requires a nomadic lifestyle. For some peoples, especially where the soils are younger and more fertile, it is possible to establish a stable village and farm the surrounding forest, rotating plots every few years but not leaving the basic area. Blue Creek in southern Belize is such a place. Sitting at the base of the limestone Maya Mountains near the Guatemalan border, Blue Creek rain forest receives about 180 inches of precipitation annually, most of it from the months of June through December. The people are Kekchi Mayan, and their ancestors farmed the Yucatán Peninsula from before the time of Tikal. The people speak their own language, plus English and Spanish. The village of about 160 people consists of several dozen wooden, dirt-floored houses with palm-thatched roofs. Small lawns are kept in front of each house, the grass cut by machete. In and around the village are planted bananas, plantain, bread-fruits, cacao, pineapple, and various citrus trees. Pigs and chickens wander around among dogs and children. Blue Creek runs through the village, and all bathing and washing of clothes is done in the creek.

Each Kekchi man clears his own *milpa*, a term for Meso-american slash and burn plots of a few acres. Sometimes villagers work together to clear a plot. Usually the milpa will be a considerable distance, often miles, from the village. Mayans are good walkers. Typical of slash and burn farming, a polyculture rather than a monoculture is planted. Instead of just one crop, several surface crops, such as corn and beans, share the same plot with root crops such as manioc, sweet potatos, and yam beans. Other crops may include some tomatoes, squash, and peppers. Studies have shown that polycultures are more resistant to insect attack, because the diversity resulting from having several crops together assures that some natural enemies of the herbivores will be present. A polyculture mimics early succession (see below, next section). Crops are planted and periodically tended until they can be harvested.

A strong sexual division of labor exists in the Kekchi village. Women keep house, tend the children, prepare food, and weave, although weaving is giving way to buying ready-made clothing. Rising well before dawn, women grind corn to prepare tortillas, bake cassava bread, and kill and dress chickens or other animal food. The men tend the milpas or go into the forest to hunt.

Kekchis use relatively modern techniques, and the men hunt with rifles, not bow and arrow.

Blue Creek represents a sort of twentieth century version of slash and burn agriculture. It is changing. North American farmers, many of them Mennonites, have moved into the village and surrounding area, bringing with them the techniques and equipment of temperate zone farming. Tractors are replacing machetes. Whether or not "modernization" will ultimately be good or bad for Kekchi lands is still under debate.

Prehistoric Intensive Agriculture

In discussing the recovery of Tikal from deforestation, I mentioned that the city, at its peak, supported 50,000 residents plus a dense sprawling suburbs. Norman Hammond (1982) has summarized archaeological evidence suggesting strongly that Mayans supported this vast population through techniques of intensive agriculture rather than mere slash and burn. There is also evidence that silviculture was practiced as well.

Intensive agriculture was accomplished by two methods, hill terracing and raised fields in swamps. Hill terracing involves the construction of stone walls along hillsides, which act to retard erosion and trap soil washed by rains. This method was widely practiced by Incas of the Andes. Mayans also utilized hill terracing, which permitted them to cultivate a given plot for much longer than ordinary slash and burn techniques because the soil fertility was preserved. Raised fields involve the excavation of drainage canals to reduce water levels and thus raise "dry" fields from what was previously swampland. Ancient Mayans not only used the raised fields for agriculture, but probably also used the canals for keeping fish and turtle, both important protein sources.

Past raised-field agriculture is revealed in part by noting patterns in aerial photographs of the landscape today. B. L. Turner and P. D. Harrison (1981) have shown that a large area in northern Belize (Pulltrouser Swamp) was under intensive wetland cultivation by Mayans between 200 B.C. and A.D. 850. Mayan techniques succeeded in preserving soil fertility and may have permitted uninterrupted farming throughout the year. Alfred H. Siemans (1982) describes extensive prehispanic agricultural use of wetlands in Belize. Using both aerial photography and remote sensing side-looking airborne radar imagery, Siemans notes that vast areas of northern Guatemala and Belize contain canals that were probably constructed by early Mayans and used in connection with agriculture. Siemans points out, however, the difficulties inherent in interpreting such important conclusions as crop yields, seasonal usage patterns, and levels of human labor on the basis of what remains today.

Arturo Gómez-Pompa and colleagues report that modern attempts to model early Mesoamerican intensive agriculture using raised fields and canal systems (called *chinampas*) can, indeed, produce impressivly high yields (Gómez-Pompa et al. 1982).

The Mayans of Tikal probably cultivated the *ramon* (or breadnut) tree (*Brosimum alicastrum*). This tree is abundant throughout the Guatemalan Peten region, and a single tree has the potential to yield 2200 pounds of edible nutritious seeds (Hammond 1982). In addition, the breadnut tree is tolerant of many soil types and grows rapidly, an ideal tree for cultivation. Its fruits and seeds would have provided sources of nutrition for both humans and domestic animals, its leaves used for animal forage, and its wood used for construction (Gómez-Pompa et al. 1982). The present abundance of *ramon* throughout areas formerly densely populated by Mayan civilization is believed by many researchers to have resulted from Mayan silviculture. Evidence cited by Hammond suggests that Mayans preserved these seeds in underground chambers called *chultunobs*. Breadnuts probably served as a "famine food," to be used when times were difficult. Pre-Columbian Mesoamerican peoples developed a variety of techniques to greatly increase agricultural yields.

Central American Successional Crop System

Slash and burn agriculture has much in common with ecological succession in that it uses the successional process to restore the soil after use for farming. Robert D. Hart (1980), working in Costa Rica, has suggested that farming can be even more analogous to succession. He presents a scheme whereby crops are rotated into and out of plots on the basis of their successional characteristics. Using such a system, Hart claims that it would be possible to utilize a plot of land continuously and productively for at least fifty years or more.

To quote Hart:

> Early successional dominance of grasses and legumes can be assumed to be analogous to maize (*Zea mays*) and common bean (*Phaseolus vulgaris*) mixtures. Euphorbiaceae, an important family in pioneer stages of early succession, can be represented by cassava (*Manihot esculenta*), a root crop in the same family. In a similar replacement, banana (*Musa sapientum*) can be substituted for *Heliconia* spp. The Palmae family can be represented by coconut (*Cocos nucifera*). Cacao (*Theobroma cacao*) is a shade-demanding crop that can be combined with rubber (*Hevea brasiliensis*) and valuable lumber crops such as *Cordia* spp., *Swietenia* spp., or other economically valuable members of the Meliaceae family to form a mixed perennial climax.

John Ewel and colleagues (1982) compared various monoculture crops with mixed species plots. They learned that the more diverse plots had significantly more root surface area and concluded that increased species diversity may enhance an ecosystem's ability to capture nutrients. They conclude, "The maintenance of root systems having high surface area of absorbing roots well distributed in the soil profile may be one of the most important

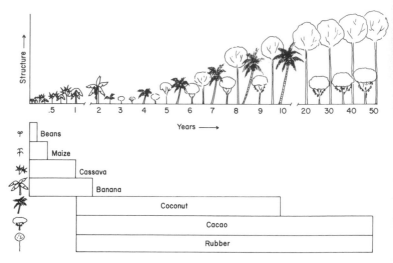

Sequence of crops that would mimic successional patterns.
See text for explanation. From R. D. Hart (1980). Reproduced
with permission.

features to strive for in designing agroecosystems appropriate for the humid tropics, where soil-nutrient storage
is often low and leaching rates are high. Such root systems can be achieved by designing systems that are diverse and long-lived."

Major Tropical Crops

Two of the three most important of the world's crops,
corn and rice, are from the tropics. They are important
carbohydrate sources, and both are well-known members
of the grass family. The third major crop that feeds the
world's peoples, wheat, is a temperate grass.

Corn, or maize (*Zea mays*), is farmed virtually everywhere in the American tropics. It is undoubtedly native
to the Americas, and evidence of its ancestry in southern
Mexico has been found. An annual, it must be replanted

every year. Corn is used to make tortillas, baked pancake-like patties, which are a staple food.

Rice (*Oryza sativa*) originated in Asia and was brought to the American tropics by the Spanish. Both upland rice and paddy rice are grown in the various places depending upon site conditions. Wild rice grows along the Amazon flood plain (varzea). Rice grains are cooked and eaten as is or sometimes ground up and made into a bread. Rice can also be fermented for wine.

Beans (*Phaseolus vulgaris*) are the most important vegetable protein source. Several varieties are grown throughout the tropics. Rice and beans are a ubiquitous dish, served in every home and restaurant. It is essentially not possible to visit anywhere in the tropics without partaking of rice and beans. (Should you be seeking variety, order beans and rice.) When rice is cooked in coconut milk and the rice and beans spiced with hot pepper sauce, the combination is delicious. In areas where corn is more common than rice, the staple is tortillas and beans.

Manioc (*Manihot esculenta*) is also referred to as cassava or yuca. It is a tuber crop whose root is very rich in carbohydrate. A perennial, manioc originated in the American tropics and will grow annually without replanting. It is an important staple throughout the Neotropics, usually ground into a paste and made into hard bread. On nutrient-poor soils, manioc tends to be loaded with prussic acid, a cyanide-containing compound that protects the root from herbivores and that must be removed prior to eating (Hansen 1983b) (page 207).

Other important vegetable crops derived from the tropics include tomatoes (*Lycopersicon esculentum*), squash (*Cucurbita* spp.), peppers (*Capsicum* spp.), and potatoes (*Solanum tuberosum*).

Cacao is used to make chocolate and comes from the cacao tree, *Theobroma cacao*. The Latin name means "food of the gods," a reference to the belief that the gods had been the original suppliers of the delicacy (Hansen 1983a). Cacao is a small tree, normally part of the rain forest understory. It periodically drops its shiny oblong leaves and grows new leaves, which are initially reddish. A cauliflorous tree, the flowers (and therefore the fruits) grow directly out of the trunk. Pollination is accomplished by a small midge that breeds on the surface of open water. Ancient Mayans raised cacao on raised fields surrounded by canals. The pollinating midge thrived in such areas. Cacao fruits can be produced throughout the year but are most abundant at the end of wet season. Fruits are highly variable in color and texture. Mammals are the principal seed dispersers. People first dry and then roast the seeds to make cocoa.

Coffee (*Coffea arabica*) is an understory shrub with many widely spreading horizontal branches. A single plant can be covered with white flowers. Wind-pollinated, the plant often self-fertilizes. Leaves are very shiny dark green. Though coffee can be grown throughout the tropics, the best beans are from regions at midelevation, between 800 and 1500 feet, that have a dry season (Boucher 1983). Coffee is often grown in large plantations with tall trees to provide shade, though there is a new coffee variety, already planted heavily in Costa Rica, that grows well in full sunlight.

Sugar Cane (*Saccharum officinarum*) is throught to have originated in New Guinea. A tall perennial grass, it is grown principally as an export crop in the American tropics. Often grown in a large-scale monoculture, it is subject to many diseases as well as insect and rodent pests.

Bananas had their origins in India, Burma, and the Philippines, developing through a series of hybridizations from two originally wild species (*Musa acuminata* and *M. balbisiana*) (Vandermeer 1983b). The first bananas in the American tropics were introduced in 1516 in Hispanola. The odd drooping arrowhead flower and the immense elongate leaves make the entire plant as distinctive as its fruit. Cultivated bananas, because they are hybrids, must be artificially propagated and cannot pollinate themselves. The tiny black spots inside the fruit, which many people believe to be seeds, are actually aborted eggs. Bananas are often grown in large scale plantations, but most

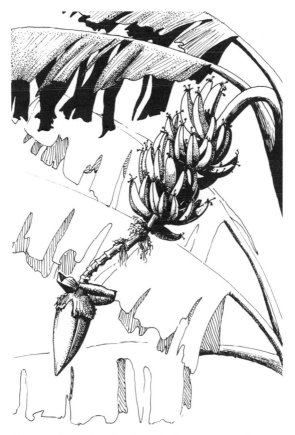

Banana plant, showing large inflorescence and
developing fruits.

slash and burn farmers have a few plants around their
houses. Eaten raw or fried as plantains, bananas are very
important to tropical peoples.

Papaya (*Carica papaya*), like cacao, is a small tree that
produces cauliflorous fruits. An early successional species, the tree has very soft wood and a brief life cycle, but
the domesticated species has never been found growing

in the wild. Its origins are thought to have been from southern Mexico or Costa Rica. Pollination probably occurs by wind, though insects may also be involved. The fruits turn from green to yellow to orange when ripe and are a favorite food throughout the tropics.

Pineapple (*Ananas comosus*) is a terrestrial bromeliad that had its origins in South America. The spikey, sharply spined leaves protect the plant, whose single flower cluster grows in the center of the leaf rosette. Domestic pineapples must be artificially propagated, though some pollination by insects can occur. Most Indian families have a few pineapples as part of their "dessert" crops.

Mango (*Mangifera indica*) is a native of Indian and Burmese rain forests that was brought either by the Spanish or Portuguese to the New World. It usually grows as a small tree with leathery, waxy green leaves that are produced in clumps toward the branch tips. The tiny flowers, which number up to 5000 on a single branch tip, are yellowish to pinkish, fragrant, and attract flies as pollinators. The fruit is variable with respect to both its size and color (ranging from yellowish to green to reddish). Mango fruits abound with terpenoids (chapter 4), which give the fruit its unique flavor.

Like Mango, the breadfruit tree (*Artocarpus altilis*) is not native to the American tropics, though it is very widely propagated and is a common sight throughout the region. A striking tree in appearance, the breadfruit has large, deeply lobed dark green leaves. The green knobbyskinned fruit, which gives the tree its name, is very large, 6–8 inches in diameter, and weighs about 6 pounds. Very rich in carbohydrate, the fruit can be fried, boiled, or baked. An infamous tree historically, the breadfruit was first discovered in Polynesia by Captain Cook. The botanist Joseph Banks, who accompanied Cook, judged it a suitable source of starch for British slaves in the Caribbean. Captain William Bligh was ordered to transport healthy breadfruits from Tahiti to Jamaica, but the wellknown mutiny on the *Bounty* prevented the success of Bligh's mission. Undaunted, Bligh returned on a second

Breadfruits.

mission and succeeded, bringing 1200 trees to Port Royal aboard the *Providence* in 1793 (Oster and Oster 1985). Initially the fruits served merely as a food source for pigs, and it was only after the emancipation of the slaves in 1838 that the people found breadfruits to be respectable as human food.

Several species of palms are important in tropical agriculture. Without doubt the best known is the coconut palm (*Cocos nicifera*), which is grown throughout the tropics. It is difficult to pinpoint the coconut palm's place of origin, but it is probably not the western hemisphere (Vandemeer 1983c). The tree was not widely distributed in the Americas before Columbus arrived. The coconut tree is unmistakable with its spreading fronds and clusters of brown coconuts below. It is fun to watch a boy climb the tree, shake loose a ripe coconut or two, and adeptly remove the husk with his machete, making a small hole so you can drink the "milk." After drinking, the coconut is shattered, and its white "meat" is eaten. Coconut can grow in any wet tropical area in almost all soil types and is by no means confined to coastal beaches. In addition to growing wild, coconut is an important plantation species, where it is grown for *copra*, made from the dried seed. The copra contains 60–68% oil, which is extracted and widely exported.

Another important plantation species is African oil palm (*Elaeis guineensis*), which was brought to the Neotropics during the slave trade (Vandermeer 1983a). This tree is widely grown in plantations throughout the Neotropics, and acreage devoted to producing oil palms is increasing rapidly. Each seed contains between 35–60% oil, the commodity for which the tree is grown.

CHAPTER 4

Evolutionary Patterns in the Tropics

BIOLOGICAL evolution is the process responsible for the way organisms look, function, and act. The arctic tundra, boreal and temperate forests, grasslands, and deserts all are comprised of plants, animals, and microbes that represent varying solutions to problems imposed upon living systems by different environments. The tropical rain forest has long been assumed to be a "laboratory of evolution" because of its extraordinary diversity of species and the complex relationships among its members.

How Evolution Works

Natural Selection

On November 24, 1859, the Britisher Charles Robert Darwin, then in his fiftieth year, published perhaps the most important book ever written on the subject of biology. This work, which bore the title *On The Origin Of Species By Means Of Natural Selection Or The Preservation Of Favoured Races In The Struggle For Life,* has since become known simply as *On the Origin of Species* (Darwin 1859).

In *On the Origin Of Species* Darwin argued that biological evolution has occurred throughout the history of life on earth and that the forms of life existent today share common ancestors with those that have lived in the past. Thus, species are not immutable but rather are changeable over time. In short, they can and do evolve. He further argued that it is the physical and biological environment of each species that imposes the conditions under which it must adapt to survive. Calling his theory "natural selection," and modeling it somewhat after artificial selection used in the process of plant and animal domestication, Darwin stated that all organisms tend to reproduce more offspring than can survive within the limits imposed by their environments and are therefore engaged among themselves in a "struggle for existence." Those individuals most suited to survive the particular conditions of the environment will tend to leave more offspring than those individuals less suited to a particular set of conditions. Any variant better suited to the environment will, therefore, be naturally selected and survive. Any genetic characteristic that enhances either an individual's survival or ability to reproduce tends to be passed disproportionately to the next generation. Those less adapted tend to perish. This theory was termed by Darwin the "survival of the fittest," a phrase he borrowed from the philosopher Herbert Spencer. Should an environment change, the species must adapt by changing or risk extinction.

Darwin's theory of natural selection was the first evolutionary theory to win wide acceptance in the scientific community. A simple but nonetheless elegant theory, Darwin's tenacious defender, Thomas Henry Huxley, upon hearing it for the first time, declared, "How stupid of me not to have thought of it." Natural selection continues today as the major explanation for adaptation.

Modern evolutionary theory defines natural selection as "differential reproduction among genotypes." Genes, the molecules of DNA that contain hereditary information, were quite unknown to Darwin, although he knew that traits could be inherited. As long as a population contains genetic variability among its members, natural selection can act. Individuals with genes that confer reproductive advantage will leave the most progeny, and, thus, those genes will increase relative to others in the population's "gene pool." Genetic variability originates through the random process of mutation, a sudden and unpredictable change in a gene. Mutation is nondirectional: environments do not cause or produce useful mutants. Selection can act only on whatever genetic variants are present. If some are adapted more so than others, they

will be passed in higher proportion to the next generation. Using the raw material of genetic variation, natural
selection "shapes" organisms to "fit" their environments.
Selection, unlike mutation, is not a random process, because only certain members of a population are best
suited for a given environment. In recent years both ge-
neticists and molecular biologists have established beyond
doubt that large amounts of genetic variability exist in
most populations. Thus there is ample raw material for
natural selection to act upon.

Darwin was not alone in his discovery of natural selection. Another Britisher, considerably younger than Darwin, independently developed exactly the same theory,
even to the point of using nearly identical terminology.
In 1858, Alfred Russel Wallace sent an essay to Darwin
that he had composed during a bout with fever while on
the island of Ternate in Indonesia (Wallace 1895). Imagine Darwin's chagrin when, on June 18, he opened and
read Wallace's brief essay and seemed to be reading his
own words. Darwin, however, had been at work on his
theory virtually from his return in October 1836 from the
five-year global voyage of the HMS *Beagle*. He had written unpublished essays on natural selection in 1842 and
1844. The latter essay, a longer and more detailed version of the first, formed the backbone of *On the Origin of
Species*. Darwin instructed his wife, Emma, to see to its
publication in the event of his death. Two of Darwin's
closest friends, Charles Lyell and Joseph Hooker, both
eminent scientists, knew of Darwin's work and of his
priority in claiming the discovery of natural selection. At
the joint urging of Lyell and Hooker, Darwin and Wallace
had their essays read together before the prestigeous Lin-
naean Society in London on July 1, 1858. Wallace gra-
ciously accepted Darwin's priority and insisted that Darwin be awarded the lion's share of the credit for
discovering natural selection. Darwin's book, published a
little over a year later, assured such would be the case.

Both Darwin and Wallace were strongly influenced by
their studies in the American tropics. Both made voyages

into rain forests to study natural history, and each man was astonished at the immensity, diversity, and complexity that he witnessed. Wallace spent four years exploring the Amazon and voyaging upriver along the Rio Negro. Darwin explored not only the Amazon but much of the rest of South America as well (Darwin 1906). He rode with gauchos over the Patagonian pampas, scaled the Andes, and became as deeply intrigued by the geology of the region as with the biology. He unearthed fossils of animals long extinct but that bore uncanny resemblances to living forms such as sloths and armadillos. Darwin also visited the Galápagos Archipelago, 700 miles west of Ecuador on the equator. These barren volcanic islands are named for their curious collection of giant tortoises (*galápagos* means "tortoise" in Spanish). The different variants of tortoises, finches, and mockingbirds that Darwin noticed as he visited several of the islands stimulated his thinking in ways that, once back in England, crystalized into the theory of evolution by natural selection.

Adaptation

An adaptation is any anatomical, physiological, or behavioral characteristic that can be shown to enhance either the survival or reproduction of an organism. Adaptations with a genetic basis are thought to result from the action of natural selection. Any organism may be viewed as a cluster of various adaptations. For instance, opossums as well as many monkeys of the American tropics possess a prehensile tail. Such a structure functions effectively as a fifth limb, lending security and mobility to the animal as it moves through the treetops. It is easy to see intuitively that the prehensile tail is an adaptive structure. Tailless monkeys or opossums would face a smaller lifetime reproductive success because of the added risk of falling. Monkeys also have excellent stereoscopic vision, because their eyes have widely overlapping fields of vision. This enables them to judge distances precisely, as when jumping from branch to branch.

Not all adaptations are obvious. In parts of Africa, blacks carry a gene that, when present in double dose (one inherited from the mother, one from the father), produces defective hemoglobin molecules, resulting in misshapen red blood cells called "sickle cells." These victims of sickle cell anemia usually die before reaching reproductive age. However, other individuals in the population who possess only one dose of the sickle cell gene and one dose of the normal gene for beta hemoglobin are not seriously disabled and are more resistent to malaria than others in the population, including those individuals with two normal genes for hemoglobin. Thus the sickle cell gene is adaptive (since the protozoan that causes malaria is part of this environment), but only when in a single dose. When in double dose it is lethal.

Not all traits are adaptive. Organisms represent the combined effects of thousands of genes working in concert. The body, physiology, and behavior of an organism represent various compromises imposed by the interactive effects of the genes that formed it. Is it adaptive that humans generally lack body hair? Maybe, but no generally acceptable adaptive explanation for hairlessness has been forthcoming. Our hairlessness could be a mere byproduct of our development, with no adaptive function now or ever in the past. A given trait may, however, have been adaptive in the past but not any longer. The human appendix, which probably functioned as a cecum, an additional area for digestion of tough plant fiber, is probably such as example.

An adaptation must be viewed in the context of the environment. Obviously, the sickle cell gene is nonadaptive in black populations from environments free of malaria. A beautifully adapted creature may appear awkward, a "mistake of nature," if seen out of its proper environment. Charles Waterton, a Britisher who explored the Amazon during the early nineteenth century, noted that the three-toed sloth (*Bradypus variegatus*) appears very ill-suited to survive when seen struggling over the ground. Sloths, however, do not normally wander about on the ground. They are totally arboreal, moving upside down from branch to branch. In his well-known account, *Wanderings in South America*, Waterton (1825) wrote of the sloth, "This singular animal is destined by nature to be produced, to live and to die in the trees; and to do justice to him, naturalists must examine him in this his upper element." Waterton then went on to describe in detail how well adapted the sloth is for its arboreal life.

Cryptic Coloration—The Art of Disguise

Cryptic coloration, or crypsis, is nature's camouflage. Thousands of species of tropical insects, spiders, birds, mammals, and reptiles exhibit cryptic coloration to various degrees. Any of these creatures, if removed from its environment, would appear obvious. For instance, the common boa constrictor (*Boa constrictor*), if placed on a white table, would be seen as boldly patterned and complexly colored in browns and golds, with stripes, diamonds, and other markings. Place the snake on the rain forest floor, among the decomposing leaves and sunflecks, and watch it melt into the background. A seven-foot serpent can seemingly disappear before your eyes. Place your hand on a tree trunk. Part of the bark can suddenly erupt in flight as a blue morpho butterfly (*Morpho* spp.), resting on the tree trunk, flies off at the disturbance. Once in flight, the deep blue wings make it easy to see. At rest, however, the blue is covered by wings virtually indistinguishable from bark. Most remarkable is the Colombian butterfly *Anaea archidona*, whose wings look exactly like dried leaves, even to the point of including lighter areas simulating insect damage and leaf skeletal patterning.

Cryptic coloration functions widely to protect animals from detection by predators. In the tropics, what appears at first to be a twig may be a walkingstick, dried leaves may be a coiled fer-de-lance, a green vine may be an emerald tree boa or vine snake, bark may be a butterfly or moth, a green leaf may be a katydid, and a tree stump may be a bird. The common potoo (*Nyctibius griseus*), a large nocturnal bird, spends its days sitting utterly still in plain sight. The bird so much resembles the end of a thick branch that it is easily overlooked. It even postures its body in such a manner as to more closely resemble a branch.

Like birds in the temperate zone, many tropical bird species have colorful males and more cryptically colored females. The little manakins, which will be discussed in chapter 6 (page 245), demonstrate this form of sexual dimorphism. Most male manakins are very colorful and easy to spot even in the rain forest interior, but females are universally olive or dull yellow and are much more

Common potoo perched atop a tree snag. The potoo is cryptic, closely resembling the dead branch.

easily overlooked. It is only the female that builds and attends the nest. The female's cryptic coloration is undoubtedly adaptive in keeping the nest more secure from discovery by potential predators.

Tropical cats also demonstrate cryptic coloration. The spotting and/or banding patterns of the ocelot, margay, and jaguar, so obvious when the animals are seen in zoos, help disguise them as they move stealthfully through the rain forest interior. The coat patterns, when seen in the shaded, sun-flecked forest interior, tend to break up the animal's outline, rendering it less visible. Although cats are predators, cryptic coloration is no less an advantage to them, as it aids them in moving undetected toward their prey.

The three-toed sloth, the animal that so impressed Charles Waterton, is less visible in the treetops by virture of the fact that algae grow in grooves in the sloth's hair, giving a distinct greenish tinge to the animal.

The first unambiguous demonstration of natural selection in action involved cryptic coloration of moths in Great Britain. The well-known example of industrial melanism, where over time moths with light-colored wings (mostly of the species *Biston bitularia*) were replaced by dark or melanistic individuals, was caused by soot from industrial pollution changing the background pattern of the tree trunks where the moths spent the day. Darkened trunks rendered light-winged moths more visible to birds, their principal predators. Dark-winged moths, once quite visible because of their melanistic pattern, were now much less visible and, thus, more fit in the polluted woodlands. Melanistic moths soon predominated in polluted areas. When pollution was reduced, light-winged moths increased again (Kettlewell 1973; Bishop and Cook 1975).

Warning Coloration—Don't Tread On Me!

Although many animals survive by camouflage, some seem to send out exactly the opposite message. Certain tropical butterflies, caterpillars, snakes, and frogs are brilliantly colored and stand out dramatically in any environment. Among the most strikingly colored of any animals are the small frogs of the family Dendrobatidae, the so-called dart-poison frogs. Bold, striped patterns of orange, red, or yellow glow almost neonlike against a black background. These diminutive little frogs live in Central and northern South American rain forests. The Choco Indian tribes of western Colombia use compounds called ba

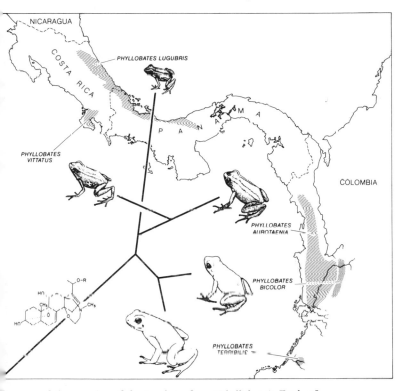

.anges of the species of dart-poison frog (*Phyllabates*). Each of
.ese species is brightly colored and conspicuous, an example
f warning coloration. From Maxon and Myers (1985).
.eproduced with permission.

trachotoxins from the frogs' skins as potent arrow poison (Maxon and Myers 1985). The fact that the skin is very toxic and the fact that the frogs are very colorful are probably not unrelated. These frogs represent a case of *aposematic*, or warning coloration. Their bold patterning serves as an easy-to-remember warning to potential predators to leave this animal alone.

Coral snakes represent another example of boldly patterned and colorful animals that are also extremely toxic. Coral snakes tend to be unaggressive unless threatened. Even though deadly, a coral snake could still suffer harm if attacked. However, its well-defined red, black, and yellow pattern is easy to recognize and avoid. Both the turquoise-browed motmot (*Eumomota superciliosa*) and the great kiskadee flycatcher (*Pitangus sulfuratus*) have been shown to instinctively avoid coral snakelike patterns (Smith 1975, 1977). Certain nonpoisonous snakes, such as some kingsnakes, closely resemble coral snakes and are therefore thought to be coral snake mimics (Greene and McDiarmid 1981). Even caterpillars of the false sphinx moth (*Pseudosphinx tetrio*) have patterning similar to coral snakes (Janzen 1980a). Mimicry in snakes and butterflies is discussed in chapter 5.

Many butterflies are among the most obvious and colorful tropical animals. Bite into a gorgeous butterfly or caterpillar, however, and prepare to be ill. Some species store compounds ingested from plants that make them toxic to birds and other predators. Those that are nontoxic often mimic toxic species (see chapter 5). Some colorful caterpillars are covered with toxic hairs.

Warning coloration is a visual manifestation of evolutionary chemical warfare in the tropics. If an animal is able to acquire or synthesize a noxious chemical it confers little protection unless potential predators can learn and remember to avoid the poisonous species. Thus most visible and boldly patterned species seem to be sending an easily memorized signal that they are dangerous. The patterns and brilliant colors warn predators to stay away.

TROPICAL SELECTION PRESSURES

Factors in the environment of an organism that influence the probability of the organism's survival or reproductive success are called *selection pressures*. Selection pressures may come from climate or physical disturbance, in the form of seasonal stresses, fire, or hurricane, or from the soil in the form of nutrient shortage or imbalance. These are *abiotic* selection pressures, since they are generated by the physical environment, not by organisms. Selection pressures may also come from competing individuals of the same species or other species, predators, parasites, disease organisms, mutualistic partners, or food species. These are *biotic* selection pressures, since they are imposed by other organisms.

Usually, combinations of selection pressures are at work. A population of howler monkeys may be ravaged by yellow fever, making the animals easier for predators such as jaguars or harpy eagles to capture. Selection pressures may also vary seasonally. The dry season often imposes significant stresses, especially in savanna areas, through both water shortage and increased likelihood of fire.

To survive, an organism must be reasonably well adapted to all of the selection pressures it is likely to encounter during its life cycle. Those individuals within the species best able to cope with the constellation of selection pressures imposed by the environment will tend to leave the most offspring. This is the essence of Darwin's and Wallace's theory of natural selection.

Abiotic Selection Pressures

Generally speaking, the climate of a tropical rain forest is more conducive to living things than other terrestrial climates. Although there is usually a pronounced dry season (chapter 1), there is no period of frost or snowfall, when water is in solid form and is unavailable as a medium for cellular growth and development. Precipitation falls as rain twelve months a year. Temperature is relatively high and nearly constant. Leaves are almost always available to herbivores and, though some individual trees may drop all leaves, there is no season in which photosynthesis shuts down entirely. Flowers and fruits are, likewise, a constant resource in a tropical rain forest. Life tends to prosper under such conditions. This is not to say, however, that the tropical climate imposes no selection pressures. Hurricanes may blow, rainfall may leach nutrients, and heat stress occurs. Yet overall, the tropical climate is far more equitable and gentle than temperate, polar, and desert climates. Physiologically, it is easier to live in the tropics than out of the tropics.

Tropical soils may impose significant selection pressures. As discussed in chapter 3, soils may be very poor in nutrient quality, and tropical plants have evolved various adaptations helping them adapt.

Biotic Selection Pressures

The tropics abound with species that exert biotic selection pressures on each other. For instance, the least favorable place for a seed to fall may be beneath its parent plant. Such proximity to an already established individual could doom the seed to be discovered and eaten by some seed predator attracted to the fruiting parent or, should it sprout, to be outcompeted by its own parent (Janzen 1971a).

Many, if not most, tropical species are probably subject to significant levels of competition from other species. The large number of species in almost every taxa in the tropics (see following section) suggests strongly that the effects exerted by the presence of other species with similar ecological needs imposes a substantial selection pressure. Interspecific competition may be one of the more important influences shaping the patterns of species diversity in the tropics. Because many different species can each exert a small but not insignificant competitive selection pressure on any given species, the term *diffuse competition* has been invented to describe the cumulative effect of competitive selection pressures imposed by other species (MacArthur 1972). Diffuse competition is undoubtedly more common in the tropics than elsewhere.

Predators represent yet another significant source of biological selection pressures in the tropics. Parasites and pathogens may also be considered here as they, too, represent a form of predation. If a tropical plant or animal manages to survive in competition with others of its own species as well as other competing species, it could still fall to a predator (or herbivore in the case of plants) or pathogen. For instance, birds seem to suffer from high levels of nest predation in the tropics, higher, in fact, than in the temperate zone (Lill 1974). Many species are secretive nesters, the females making few daily trips to feed nestlings and nestlings being less vocal than their temperate zone counterparts. If one begins to add up potential nest predators, the list soon becomes staggering. Toucans, forest falcons, snakes, monkeys, kinkajous, cats, even army ants may attack bird nests! A study of the nesting behavior of the turquoise-browed motmot in the Mexican Yucatán revealed the importance of predation. Motmot nests were located in the walls of Mayan ruins. Approximately 36% of the nests studied were terminated by predators, and nest location was an important factor in protection from possible predation (Scott and Martin 1983).

It has been suggested that one reason tropical bird species lay fewer eggs per clutch than temperate species is because the attention created by attending a larger brood would increase the risk of attracting predators to the point that most broods would be discovered and devoured (Ricklefs 1969a, 1969b, 1970). Smaller clutches are more adaptive because they can be raised more secretly. In other words, because of predators, the probability of raising two chicks is substantially higher than raising three, even if food is abundant. Food, however, may not in actuality be abundant for tropical nesting birds, and food shortage, an alternate explanation to account for low clutch size in the tropics, has also been suggested (Lack 1966).

Frogs represent another of many possible examples of

the importance of predation as a tropical selection pressure. Male frogs must call at night to attract females. Lucky males then get to mate. Unlucky males, however, attract bats that are specialized to capture frogs by homing in on their amorous calls. Working on Barro Colorado Island in Panama, Merlin D. Tuttle and Michael J. Ryan (1981) studied the fringe-lipped bat, *Trachops cirrhosus*, and found that this species responds to a wide variety of frog calls. The more frequently or longer a frog calls, the more likely it is to attract a bat. Bats also differentiated between poisonous and nonpoisonous frogs, strongly preferring to attack nonpoisonous animals. Male frogs risk much for love.

Even something as simple as preparing for a trip to the tropics reflects the importance of biotic versus abiotic selection pressures. If you were going to the arctic, you would spend time and dollars on warm clothing, snowshoes, and various other paraphernalia designed to protect against the potentially harsh elements. For a human visitor to "adapt" to the arctic requires profound respect for the climate (though mosquitos can be an awesome nuisance in the arctic summer). Preparing for a trip to the tropics, in contrast, requires a trip to the doctor for the various shots and pills that protect against the pathogens and parasites that await the human tropical visitor.

My point is that the presence of large numbers of organisms means that most important selection pressures in the tropics tend to be biotic. When we speak of interesting adaptations of tropical species, we most often speak of adaptations involving interactions among species. Most organisms, physiologically, can live easily in the warm and wet tropical climate. In the tropics, evolution has been most influenced by interactions among the myriads of plants, animals, and microbes that occupy it.

DIVERSITY GRADIENTS

High species diversity is the rule in the tropics. With few exceptions, most major taxa, the flowering plants, ferns, mammals, birds, reptiles, amphibians, fresh water fish, insects, spiders, and snails, all exhibit their highest diversities in the tropics. Why do tropical ecosystems contain so many species?

Charles Darwin was impressed by high tropical species diversity. Darwin realized that diversity declines latitudinally as one travels north or south from the equator, a point he noted in chapter 3 of *On the Origin of Species*. The reduction in diversity with increasing latitude has been noted often since Darwin and is now termed a *latitudinal diversity gradient* (Connell and Orias 1964; MacArthur 1965; Pianka 1966). Dobzhansky, in his important 1950 paper titled "Evolution in the Tropics," noted that only 56 species of breeding birds occurred in Greenland, while New York had 195 breeding species. However, Guatemala had 469, Panama 1100, and Colombia 1395! Breeding bird diversity increased almost twenty-five times from Greenland in the arctic to Colombia on the equator. For snakes, Dobzhansky noted that 22 species occurred in all of Canada, whereas 210 were found in Brazilian forests and savannas.

Dobzhansky (1950) offered a series of questions about latitudinal diversity:

> Since the animals and plants which exist in the world are products of the evolutionary development of living matter, any differences between tropical and temperate organisms must be the outcome of differences in evolutionary patterns. What causes have brought about the greater richness and variety of the tropical faunas and floras, compared to faunas and floras of temperate and, especially, of cold lands? How does life in tropical environments influence the evolutionary potentialities of the inhabitants?

Dobzhansky believed that part of the answer to high tropical diversity rested in the benign nature of the tropical climate. He argued that polar and even temperate climates impose such significant physical stresses that fewer organisms have been able to adapt over evolutionary time. The tropics, in contrast, offer a more equitable climate, permitting speciation with little extinction. Specialization among organisms has resulted in high diversity as more and more species have "packed" into the tropics over time. Dobzhansky's hypothesis was highly speculative, but it did much to stimulate thinking about latitudinal diversity gradients.

Tropical moist forests, not the tropics in general, are the source of the high diversity. Tropical savannas, grasslands, and deserts do not exhibit substantially higher diversities of plants and animals than their temperate counterparts. It is the tropical lowland and montane forests, the rain and cloud forests, that are so extraordinarily diverse.

Among the various taxonomic groups experiencing maximum diversity in equatorial forests, mammals represent an unusual example because most of the increase in mammalian diversity from temperate to tropical regions is due to bats. Although monkeys, sloths, anteaters, and various marsupials all contribute to the enhanced diversity of the tropics, bats add by far the most species. Bats will be discussed later in this chapter.

Diversity has several components, all of which are high in tropical forests (Whittaker 1975; Boulière 1983). First, there is *within habitat* diversity. This is the number of species in a given area of forest. Tropical forests are utterly cluttered with species. No ecosystems are known with higher within habitat diversities. Second, there is *between habitat* diversity, or the change in species composition from one habitat to another, similar habitat. Tropical forests rank very high in this diversity component also. This means that one can find somewhat different species of hummingbirds, tanagers, antbirds, etc. as one moves from one forest site to another. Finally, there is regional diversity, the total number of species found in all habitats over a large geographic region. Again, the tropics exceed all other regions in total diversity.

Several explanations have been posited to explain how tropical forests have come to be so diverse (Connell and

Orias 1964; MacArthur 1965; Pianka 1966). No simple or single explanation has yet emerged. Causal factors for species diversity are difficult to test in a way that limits the test to only one factor, excluding all other possiblilites. Below I review the major hypotheses for high tropical forest diversity. None may be correct, or all may contain some element of truth.

The Stability-Time Hypothesis

It has been argued that the tropics have been around for such a long time that large numbers of species have had time to evolve. This has been coupled with the argument that the tropics are climatically stable, thus permitting high speciation with low extinction, resulting in an accumulation of species. Unfortunately, there is no obvious way to test this idea since it is basically historical.

The tropics do seem to serve as a refuge for certain groups poorly represented outside of the tropics. Tree ferns, orchids, bromeliads, reptiles, and amphibians are dramatically more diverse in the tropics. In the case of reptiles and amphibians, the tropical warm, wet climate probably does contribute toward their species diversity. Reptiles and amphibians are ectothermic, unable to physiologically regulate their body temperatures. Amphibians require moisture to keep their skins wet. The tropics seems climatically far better suited for their needs than temperate or polar regions.

The Ice Age undoubtedly affected polar and north temperate regions more dramatically than it did the equatorial region. It is a mistake, however, to conclude that the tropics were unaffected by the dramatic movement of the northern glaciers. Indeed, as I shall discuss later in this chapter, the disturbances caused by the Ice Age may have provided a great stimulus to speciation in the tropics.

The argument that high tropical diversity is caused by the region being old and climatically gentle is not seen as a sufficient explanation.

The Interspecific Competition Hypothesis

One of the most difficult measures to make in ecological research is the degree to which competition occurs between two or more species. One must be able to demonstrate that two (or more) species are seeking the same limited resource, and then measure the degree to which each species affects the other. It is essential to identify the resource being contested and demonstrate the fact that it is a *limited* resource. If it were not limited, there would be enough for each species, thus, no competition. It is also essential to show how the competition affects each species. Does one outcompete the other? Do they somehow divide the resource? Does one expand its population in the absence of the other?

One prevalent hypothesis to explain high species diversity in the tropics argues that high levels of competition among species have resulted in increased specialization. Each species becomes a specialist focusing on a specific resource that it and it alone is best at procuring. This trend toward specialization due to interspecific competition leads to the "packing" of greater and greater numbers of species into tropical ecosystems while at the same time reducing the intensity of competition among species as each specializes to its exclusive pool of resources. Ecologists describe the total constellation of resources required by a given species to be that species' *ecological niche*. The interspecific competition hypothesis argues that niches are narrower in the tropics than in the temperate zone because competition has compressed them. Note that there is a strong historical component to this hypothesis. Though competition may have exerted a major influence in the past, now that specialization and niche compression have occurred, competition may be quite minimal or even nonexistent. This is a very difficut hypothesis to test.

Evidence for this hypothesis is mostly circumstantial. Few direct demonstrations of interspecific competition or its effects have been found in the tropics. Certain patterns suggest, however, that competition among species has been an influential component of tropical evolution. Varying bill shapes and gradations in body sizes within many bird groups (see this chapter, page 143, and chapter 6, pages 226, 267) suggest that competition has influenced the evolutionary history of these groups. Different body sizes and bill characteristics usually reflect specialization for capturing differing food items. Though the act of food capture *per se* could and probably does select for such specializations, it is quite likely that the presence of similar species with similar ecological needs could act as a strong selection pressure in producing divergence among species. Insectivorous bats feed on different sized prey items and also forage at different heights in the rain forest, a pattern possibly reflective of competition among the bat species.

The fact that clusters of similar species develop differences in foraging areas is an indirect indication that competition has been at work. In a study of four antbird species of the genus *Myrmotherula*, it was found that each species foraged at a preferred height above the ground and that foraging heights had relatively little overlap (Terborgh, in MacArthur 1972). This suggests that each species is being restrained by the others from expanding its foraging height niche. Thomas W. Sherry (1984) recently described a similar result for a group of three flat-billed flycatchers in Costa Rica. Each species forages at a different height in the forest. Sherry also noted that certain flatbill species replace others in specific habitats. One species occurs in forest, another very similar species only in successional areas. John Terborgh and John S. Weske (1975) studied Andean bird communities and found evidence for interspecific competition. They studied two Andean mountain peaks in Peru, one of which was quite isolated. Because of its isolation, colonization by birds was less frequent, and thus this peak had a much reduced bird species diversity. Terborgh and Weske estimated that

80–82% of the bird species that would have occupied the isolated peak, had it been part of the main body of the Andes, were missing. However, of the species that did occur, 71% had expanded their altitudinal range (compared with the other more diverse mountain peak) in the absence of competitors. In the absence of similar species, a given species tended strongly to expand its range on the mountainside. Terborgh and Weske concluded that the combined effects of direct and diffuse competition account for approximately two-thirds of the distributional limits of Andean mountain-dwelling birds.

Interspecific competition exists among flowering plants for pollinators and seed dispersers. In Trinidad's Arima Valley, David W. Snow (1966) found that eighteen species of the shrub *Miconia* had flowering times that were staggered in such a way that only a few species were flowering in any given month. He hypothesized that the staggering is an evolutionary result of competition among the *Miconia* for birds that eat the fruit and disperse the seeds. Any *Miconia* species that flowered when most others didn't would be able to attract more birds to disperse its seeds, thus it would have a selective advantage compared with others of its own as well as other species. Over time, the staggered flowering pattern emerged. E. Raymond Heithaus and his colleagues (1975), studying bats in Costa Rica, observed a similar pattern among plants that are bat pollinated. Of twenty-five commonly visited plant species, an average of only 35.3% flowered in any given month. Also in Costa Rica, F. Gary Stiles (1977) noted that flowering peaks of hummingbird-pollinated plants were staggered and that they shifted somewhat from one year to another, suggesting that seasonal and annual variations in rainfall are also important in influencing flowering peak sequence.

There is no doubt that specialization is a widespread phenomenon in the tropics. But the degree to which direct or diffuse competition with other cohabiting species was and continues to be the cause of the specialization is unclear. Specialization could be largely the result of adaptation to food resources, quite apart from any influence of competition. It is probably true, however, that competition among species has figured substantially and continues to contribute to diversity patterns in the tropics.

The Predation Hypothesis

One result of intensive competition can be the extinction of one or more of the competing species. What can prevent such extinction? One factor is predators. Suppose four caterpillar species are competing for the same plant. One species begins to win, and the others are being driven toward extinction. What was a four species system is about to become a one species system. But suppose birds prey on the caterpillars. Which of the four species are the birds likely to take? The most obvious and abundant species would seem the likely choice. The result of predation by the birds would be to reduce the growing population of the "winning" species, allowing the other "losing" species to regain some control of the resources and increase in population. This scenario describes the predation hypothesis of diversity.

The argument of the predation hypothesis is that predators prevent prey species from competing within their ranks to the point where extinction occurs. Predators constantly switch their attentions to the most abundant prey, thus the rarer the species, the safer it is from predators. This idea is termed *frequency-dependent selection* because the intensity of selection (in the form of predation) depends directly on the abundance of the prey. The result of predator pressure is to preserve diversity by preventing extinction by competition.

Note that the predation hypothesis is basically opposite from the interspecific competition hypothesis. The competition hypothesis says that competition among species promotes diversity by leading to specialization, narrower niches, and tighter species packing. The predation hypothesis says that predators minimize interspecific competition, thus permitting coexistence among competing species. The predation hypothesis does not predict extreme specialization. Indeed, specialization would be less likely to occur because predators keep competition levels low. Likewise, the predator hypothesis predicts wide niche overlap among similar species.

There is little direct evidence supporting the predation hypothesis. It has been shown that birds and other predators develop *search images* of prey and, thus, do indeed focus on a specific prey type when it is abundant (Cain and Sheppard 1954). Predators often do switch their attentions according to the relative abundances of their prey species. Tropical forests do contain impressive predator densities, circumstantial evidence for the possible importance of predator effects. This in itself, however, is not sufficient proof that the predation hypothesis operates as a major process influencing species diversity in the tropics.

Both the interspecific competition and predation hypotheses could operate simultaneously in the tropics. Predators may compete among themselves and specialize on various prey groups that are prevented from competing by predation!

The Productivity-Resources Hypothesis

One frequent suggestion to explain high diversity in the tropics is that high productivity permits more species to be accommodated. The idea here is that the high tropical productivity and biomass translate into more space for more species. Daniel Janzen (1976), in a paper titled "Why Are There So Many Species of Insects?" concluded by saying, "I think that there are so many species of insects because the world contains a very large amount of harvestable productivity that is arranged in a sufficiently heterogeneous manner that it can be partitioned among a large number of populations of small organisms." Janzen was not restricting his speculation to the tropics, but his remark fits the tropics well. There is indeed a tremendous potential harvestable productivity, and there are lots of spaces for insects in the three-dimensionally complex rain and cloud forests.

Bird species diversity often correlates with foliage height complexity in the temperate zone (MacArthur and MacArthur 1961). The more layers of trees there are, the more bird species that can be accommodated. In the tropics, however, bird species diversity does not correlate well with foliage height diversity (Lovejoy 1974). This indicates that the tropical forest must be more complex than temperate forests, offering resources not measured by simple structural analysis. James Karr (1975) noted that tropical forests offer certain resources for birds found in no other ecosystem, at least not to the same degree of abundance or constancy. These "additional" tropical resources include vines, epiphytes, and dry leaf clusters (which add both space and potential food for birds), large insects and other arthropods, and the year around abundance of nectar and fruit. Karr and others have pointed out that neither nectar specialists, such as hummingbirds, nor certain fruit-eating specialists could hope to exist outside of the tropics, since they are so dependent on constant availability of nectar and/or fruit. Forest gaps also represent resources. Many species of plants and animals may be essentially "gap-specialists." Without gaps, these species would not be present.

One example of the greater range of resources available in the tropics comes from a study by Robert A. Askins (1983). Comparing the woodpecker communities of Maryland, Minnesota, and Guatemala, Askins recorded seven woodpecker species in Guatemala compared with only four in each of the temperate study areas. He found that although there were more species of tropical woodpeckers, these were in reality no more specialized in their foraging behavior than their temperate relatives (thus, their niches were not narrower in the tropics). They did utilize a wider range of resources, however. Some of the tropical species fed on termites and ants, probing into the excavations made by these insects, and the tropical species fed much more heavily on fruit than the temperate woodpeckers. Two species in particular, the black-cheeked (*Melanerpes pucherani*) and the golden-olive (*Piculus rubiginosus*), utilized resources not used by temperate species. The black-cheeked frequently fed on fruit, and the golden-olives probed moss and bromeliads.

Another example of "additional" tropical resources is provided by the work of J. V. Remsen, Jr. and T. A. Parker III (1985), who studied the bird community of bamboo stands. Remsen and Parker learned that in tall bamboo stands, where plants reach heights of up to 50 feet, as many as twenty-one bird species are specialized in some way to feed only in bamboo. Nine species specialized on eating bamboo seeds, and twelve were insect foragers. An additional sixteen species of insect foragers were found mostly in bamboo but also in other habitats.

Work by Gary H. Rosenberg (1985) revealed that ephemeral Amazonian river islands provide a resource to which birds have specialized. Rosenberg found that fifteen bird species were restricted to island habitats such as sandbar scrub and young *Cecropia* forest.

Mammalian diversity also correlates with productivity and habitat characteristics. Louise H. Emmons (1984), studying Amazonian mammals (excluding bats), learned that both the density and number of species of mammals correlated positively with soil fertility and undergrowth density. Large mammalian species tended to range widely and maintain relatively constant densities over large areas, but small species varied dramatically in numbers and diversity from one study site to another.

There seems little doubt, when examining the diversity of birds or other animal taxa, that the lushness and largeness of tropical forest is at least in part responsible for the high diversity. But in a sense this begs the question, because it does not explain why the plants are so diverse. In the Amazon Basin, there are 1700 species of trees! Why are there so many kinds of trees? I will take up this question a bit later in the chapter.

No convincing and all-encompassing theory of tropical species diversity has as yet been put forth. One major difficulty with both the interspecific competition and predation hypotheses is that it is not obvious why tropical regions should be more influenced by these processes than temperate ecosystems. Some have argued that disturbance is less of a force in the tropics, thus permitting more intensive and continuous ecological interactions. There are many, however, who doubt that disturbance is less important in the tropics (see below). As I stated earlier, there are likely to be elements of truth in each of the diversity hypotheses. Perhaps that is the real key to tropical diversity. Several forces are working simultaneously in promoting and preserving species diversity.

Mild seasonality and consequent predictability of resources are widely believed to contribute to the high degree of specialization exhibited by tropical species. In the tropics, species can "afford" to evolve as specialists because resources vary less dramatically with the seasons (i.e., something is always flowering, something is always fruiting). In the temperate zone, species must be more adapted as "generalists," since they are confronted with selecting different resources in different seasons. The fact that the tropics are frost-free could be *the* major factor responsible for the higher diversity. In a sense, being frost-free could be the "resource" of most importance.

Adaptive Radiation Patterns

The abundance of species in the tropics is far from random. Many obvious and important patterns emerge when studying tropical diversity. One such pattern is *adaptive radiation*, which occurs when one type of organism gives rise to many different species, each adapted to exploit a different set of environmental resources. Darwin discussed adaptive radiation of a group of small finches on the Galápagos Islands, a group so important to his theory of natural selection that they now bear the common name of Darwin's finches (Lack 1947). These small chunky black and brown birds vary in bill size and shape, though otherwise they appear quite similar. In his account of the voyage of the *Beagle*, Darwin wrote of these birds, "Seeing this gradation and diversity of structure in one small, intimately related group of birds, one might really fancy that from an original paucity of birds in this archipelago, one species had been taken and modified for different ends." Two excellent examples of adaptive radiation are tropical bats and tyrant flycatchers.

Bats

The basic bat is a marvel of adaptation. It is the only mammal capable of true flight (excluding those who invented and use airplanes and space shuttles). Bats appear in the fossil record as far back as the Eocene, approximately 60 million years ago (Carroll 1988). Their closest relatives are the insectivores, the moles and shrews. The most important adaptation of bats is the modification of the forelimb (arm) as a wing through elongation and enclosure of the arm and finger bones within a membrane of skin. Like birds, bats have large hearts, light body weight, and high metabolism. While most birds are diurnal, all bats are nocturnal. In terms of species diversity, birds are far more successful as a group than bats; there are approximately 8700 bird species, about ten times the number of bat species (Fenton 1983).

Bats in the American tropics are all members of the suborder Microchiroptera, the insectivorous bats. Microchiropterans capture prey on the wing and avoid obstacles by using echolocation. These animals emit loud, high-pitched vocalizations (mostly inaudible to humans) that bounce off objects of approximately the same size as the wavelength of the emitted sound, thus providing the bat with an effective sonar system for locating small nearby objects, such as flying insects. Most microchiropterans have very prominent pinnae or external ears that aid in receiving the echolocation signals.

The adaptive radiation of microchiropteran bats in the American tropics is impressive. These animals were originally adapted to feed on insects captured in the air using

A typical insectivorous
bat.

echolocation. Indeed, many insectivorous species still feed in this "traditional" way. However, there are now also fruit-eating, nectar-eating, pollen-eating, fish-eating, frog-eating, bird-eating, lizard eating, mouse-eating, bat-eating, and even blood-lapping bats (the infamous vampires). From an insectivorous ancestor (or ancestors) tropical bats have radiated into dramatically different feeding niches, taking advantage of the tremendous diversity of rain forest resources.

Daniel Janzen and D. E. Wilson (1983) studied the pattern of bat diversity in Costa Rica and tallied a total of 103 species (compared with 40 species in the entire United States). Of these, 43 are insectivorous, 25 are frugivorous, 11 nectarivorous, 2 carnivorous, 1 piscivorous (fish-feeder), 3 sanguivorous (blood feeders), and 18 feed on some combination from the above list.

Among the insectivorous bats, some capture prey by aerial foraging and some by foliage gleaning. Aerial foragers catch insects on the wing, while foliage gleaners pick their prey off leaves, branches, and even the ground.

The false vampire bat, *Vampyrum spectrum*, is a carnivore. It captures sleeping birds as well as rodents and other bats, probably locating some of its prey by olfaction. Many bats have a keenly developed sense of smell. This is one of the largest of the Neotropical bats, with a wingspread of approximately 30 inches. It has prominent ears, a long snout, a large "leaf" nose (see below), and long sharp canine teeth. Its generally ferocious appearance misled people into believing it to be a vampire, which it is not.

Flattened leaflike noses are common among fruit-eating bats as well as carnivores. The flattened nose may be an adaptation aiding the bat with echolocation. These bats typically carry large food items and must therefore

A typical leaf-nosed bat.

emit their sonar vocalizations through their noses rather than mouths. The leaf nose may help focus the signal (Fenton 1983).

The fishing bulldog bat, *Noctilio leporinus*, is another excellent example of adaptive radiation. This species uses its sonar, not to locate insects in the air, but instead to locate fish just beneath the water's surface (Brandon 1983). It gets the name "bulldog bat" from its flattened but puffy face, with small eyes, short pointed ears, and prominent cheek pouches. A large bat, its wingspread is 24 inches. It has long toes with prominent, sharp claws. Using its sonar, the bat detects small fish and crustaceans breaking the surface of calm rivers and pools. It swoops down, gaffing the fish with its large well-clawed feet. It then transfers the piscine to its mouth, where it stuffs it in its large cheek pouches as it grinds it up bones and all. I once spent a night in a motel room with a *Noctilio* that a colleague was bringing back for study in the United States. The bat fed on fish from a pail in the bathroom. It consistently devoured fish headfirst and hung contentedly upside down (feet attached to a picture frame) as it munched away, fish tail protruding from its mouth. The fact that the bat reeked of fish and strong musk, plus the constant sound of fish bones being ground up, made for fitful sleeping that evening.

The most infamous Neotropical bat is the vampire bat, *Desmodus rotundus*. This extraordinary animal feeds entirely on the blood of mammals such as tapirs and peccaries. In recent years vampires have prospered because of cattle and swine, both of which afford an easily accessible source of blood (D. C. Turner 1975). Vampires fly from their roosting caves at night to locate prey. The bat finds its victim both by olfaction and vision. Remarkably agile (for a bat), the vampire scurries over the ground on its hind legs and thumbs, climbing on to the sleeping animal. Using specialized incisors, the bat slices into the superficial skin layers and initiates bleeding. The cut is so sharp that the prey animal rarely awakens. The bat's saliva contains an anticoagulant, so the blood flows freely while the

A fishing bulldog bat capturing a fish.

bat feeds. The bat's digestive tract is modified to deal with blood, which is extraordinarily high in protein. Another vampire species, *Diaemus youngii*, feeds only on birds.

Vampires do attack people, though rarely. I have seen Mayan Indians in Belize that have been bitten about the face and fingers by vampires. Fortunately for these people, none of the bats was a carrier of rabies, though in many places vampire bats do transmit this serious disease. Vampires carrying rabies may show no symptoms themselves, a fact that makes them potentially very dangerous

A vampire bat showing walking posture.

but that also indicates a long evolutionary relationship between the vampire bat and the rabies virus.

Bats exhibit adaptive radiation not only in their feeding behaviors but also in their choice of roosting sites. Thomas H. Kunz (1982) notes that the roost is the place in which a bat spends about half its life and roosts are the sites of most social interactions, mating, rearing young, and food digestion, as well being refuges from adverse weather conditions and predators. Bats may be colonial or solitary roosters. Roost sites include caves, crevices, or hollow trees. Some bats roost in foliage, often modifying it to suit their needs. The small bat *Thyroptera tricolor* roosts in furled *Heliconia* and banana leaves, attaching to the slick leaf with adhesive disks on the legs and wrist joints of the wings. *Ectophylla alba*, a small, beautiful, all-white bat, goes one step further. It is one of several species to actually construct a tent, in this case out of a *Heliconia* leaf. The white bat forces the huge leaf to droop by carefully chewing veins that are perpendicular to the midrib. The leaf is only partially chewed and the result is a protective, thick tent where a half dozen or so of these diminutive bats can cuddle in safety (Kunz 1982).

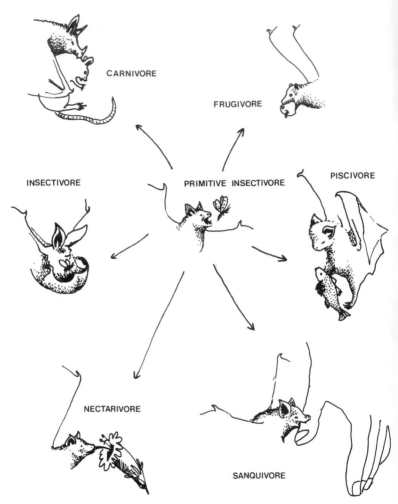

CARNIVORE

FRUGIVORE

INSECTIVORE

PRIMITIVE INSECTIVORE

PISCIVORE

NECTARIVORE

SANQUIVORE

Adaptive radiation of Neotropical bats.

Bat courtship and mating is also diverse. Some bats are monogamous, and some are highly promiscuous. In some species, male bats establish a "harem" of females. Thomas H. Kunz and his colleagues (1983) studied harem social organization of the bat *Artibeus jamaicensis* on Puerto Rico. Bat clusters in caves consisted of two to fourteen pregnant lactating females and their offspring plus a single adult male. Older and heavier (larger) males had larger harems than the younger animals, indicating that "bat men do better than bat boys" (my words, not Kunz's). Nonreproductive females and haremless males formed their own separate groups. One reason for harem formation may be that females select caves or hollow trees in which to raise young and these areas are small enough to be effectively defended by a single dominant male. Males would therefore compete against each other for access to females. A successful male would obviously enjoy a tremendous reproductive advantage.

I'll say more on bats in the section on coevolution, page 161.

Tyrant Flycatchers

One of the most diverse families of birds in the world is the tyrant flycatchers, and, thus, they provide another excellent example of the evolutionary process of adaptive radiation. There are 367 species of Tyrannidae (Austin 1985), all of which are confined to the New World and most of which are in Central and South America. Melvin A. Traylor, Jr. and John W. Fitzpatrick (1982), in a comprehensive review of tyrant flycatcher diversity, have estimated that 1 of every 10 bird species in South America is a tyrant flycatcher. They estimated that tyrant flycatchers make up 20% of the perching bird species in Colombia and 26% in Argentina. They abundantly occupy all habitats: rain forests, cloud forests, savannas, puna, and paramo. Though the name *flycatcher* is meaningful in most cases, the methods of capture and the types of insects captured vary tremendously among species. Finally,

A typical tyrannid flycatcher.

some have diverged entirely from capturing insects, becoming fruit eaters.

Traylor and Fitzpatrick attribute the immense success of the tyrannids to their versatility of feeding methods. They describe the basic feeding behavior as *sally-gleaning*, meaning that the bird sits on a perch and flies out to capture an insect, either in the air, on the ground, or on a leaf surface. Because flycatchers prefer exposed perches, they are well suited to invade any open habitat such as forest edges, riverbanks, savannas, etc. From the fundamental sally-gleaning technique has evolved many specializations. There are hawkers, ground feeders, runners, hoverers, water's edge specialists, perch-gleaners, and fruit eaters (see also Fitzpatrick 1980a, 1980b, 1985).

Body size as well as bill size and shape varies widely among the tyrannids, much more so than in the temperate zone (Schoener 1971). Among insectivorous birds in general, and tyrannids in particular, large-billed species are much more evident in the Neotropics, possibly due to greater availability of large arthropod prey (Schoener

1971). Interspecific competition within the group may also have provided selection pressures resulting in divergence of body size, bill, and feeding characteristics.

Some tyrant flycatchers have uniquely shaped bills. The northern bentbill (*Oncostoma cinereigulare*) has a short but distinctly down-curved bill. The northern royal flycatcher (*Onychorhynchus mexicanus*) has a long and flattened bill, and the tiny spadebills (genus *Platyrinchus*) have extremely wide, flattened bills.

Size and bill variation is strikingly evident in a group of species that all have yellow bellies and striped heads. Indeed, they look very much alike, except for body size and bill shape. In Panama alone, seven species occur. The great kiskadee (*Pitangus sulphuratus*) and boat-billed (*Megarhynchus pitangua*) are similar in size, but the latter has a wide, flattened bill. The lesser kiskadee (*Pitangus lictor*) looks like a small version of the greater and differs from the very similarly sized rusty-margined flycatcher (*Myiozetetes cayanensis*) by having a longer bill. The white-ringed (*Conopias parva*) and social (*Myiozetetes similis*) flycatchers differ only in minor facial characteristics, and the gray-capped (*Myiozetetes grandadensis*) has a gray cap that helps distinguish it from the others (Ridgely 1976).

Tyrannids abound in rain forests. In my Belize study area, I find 23 species, ranging from the 10-inch great kiskadee to the tiny 3.5-inch white-throated spadebill (*Platyrinchus mystaceus*). Traylor and Fitzpatrick (1982), citing a survey done by Terborgh, note that at a single site in Manu National Park in the Peruvian Amazon, 65 (23%) of the 281 species of perching birds are tyrannid flycatchers! Tyrannids are most abundant in the canopy and along the forest edge, but there are species that occur virtually anywhere from the ground up.

Some flycatchers switch their diets regularly, and others are basically opportunistic, switching according to the relative abundances of various insect prey (see predation hypothesis, above). One seasonal and dramatic switch is seen with the eastern kingbird, *Tyrannus tyrannus*. One of 32 tyrannids that migrate to breed in North

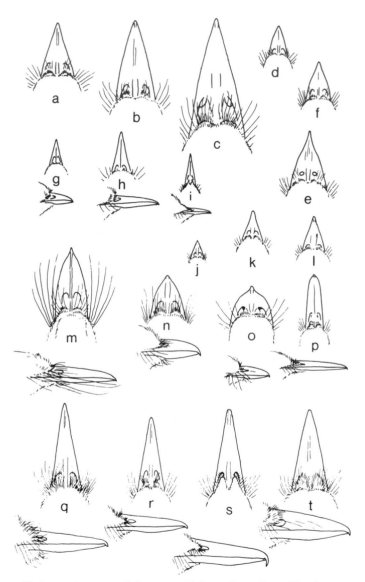

Bill shapes in some of the tyrannid flycatchers. From Traylor and Fitzpatrick (1982). Reproduced with permission.

America, the eastern kingbird feeds on insects on its summer breeding grounds. However, when on its wintering grounds in southwest Amazonia, it forms large flocks and feeds mostly on fruit, the flocks wandering nomadically in search of fruiting trees (Fitzpatrick 1980b). Another flycatcher, one which resides year-round in the Neotropics, has evolved a fruit-eating diet. The nondescript ochre-bellied flycatcher (*Pipromorpha oleaginea*) is the most abundant forest flycatcher in Trinidad (and it occurs abundantly elsewhere in the Neotropics). Its inordinate abundance compared with other tyrannids may be due to its diet shift from arthropods to fruit, which is an abundant and easily "captured" resource (Snow and Snow 1979).

Overall, the abilities of varied tyrannids to find food in virtually all habitats has likely been a major factor promoting speciation within the group. Species have special-

The head of a northern
royal flycatcher, one of the most
spectacular of the
tyrannids.

ized on certain types of arthropods and other food, captured in distinct ways.

SPECIATION IN THE TROPICS

The impressive adaptive radiations apparent in bats, tyrant flycatchers, and other groups in the tropics indicate that the process of speciation has been active. Over time one species may evolve into a different species, or into several different species. *Speciation*, the splitting of one species into two or more new species, where a single *gene pool* becomes divided into several distinct gene pools, is the process that has produced the vast species richness of the Neotropics.

How Speciation Occurs

Ernst Mayr (1963) defined a biological species as "an actually or potentially interbreeding population, reproductively isolated from other such populations." The key to identifying a species, according to Mayr, is reproductive isolation. Only members of the same species can mate and produce fertile offspring. A horse and a donkey are capable of mating, but the union results in a mule, a robust but infertile animal. Therefore, the gene pools of the horse and donkey are isolated from one another, and the horse and donkey are separate species.

The speciation process involves the establishment of reproductive isolation between two populations. In order for this to happen, the flow of genes between these populations must be prevented for a sufficient number of generations to permit genetic divergence adequate to establish reproductive isolation. Once two populations have genetically diverged, they will be unable to produce fertile offspring and, thus, will be distinct species.

One important means by which populations can become fragmented and gene flow thus interrupted is *geographic isolation*. Mountains, rivers, deserts, savannas, all represent possible barriers between rain forest sites.

Should a mountain be formed by uplifting of earth's crust, as has happened extensively in the Andes Mountains, what was once a contiguous area of forest will be fragmented by the mountain range. Individuals on one side of a mountain are prevented from mating with those on the other side because they simply cannot cross the mountain. Once populations are separated by geographic factors, there is the strong possibility of genetic divergence over time among the subpopulations. For instance, a population of lizards on the south side of a mountain may be subjected to different selection pressures than the population on the north or east side of the mountain. The result of this selection pressure difference is that each population will be selected for different characteristics, and thus for different genes. Also, small isolated populations are subject to chance factors affecting their gene pool, a process called genetic drift. (For example, an isolated population of, say, twenty individuals, where only one of the twenty contains gene "A," could lose gene A very easily if the individual carrying it either fails to mate or does mate but does not pass on that particular gene.) Geographic isolation helps promote speciation, because it allows the buildup of genetic differences between populations by preventing gene flow among them. Speciation is promoted when populations fragment or when a small subset of a population "breaks off" from the parent population and becomes the founder of a new population in a different location. This was undoubtedly the case for Darwin's finches. A small flock (or perhaps a single pregnant female) flew (probably "helped" by a storm) from mainland South America to a landfall on one of the Galápagos Islands, becoming the founding population of the Darwin's finches.

The Effects of Topography

Central and western South America are both geologically active areas. The Andes mountain chain is responsible for creating a diversified complex of habitats as well as providing numerous climatic and physical barriers that greatly enhance geographic isolation among populations. The massive Amazon River and its numerous wide tributaries have also served to isolate tracts of forest and savanna. The width of the Amazon and its tributaries is sufficient to isolate populations of birds whose individual members are reluctant to cross such a wide expanse of water (Haffer 1985; Haffer and Fitzpatrick 1985). For example, blue-crowned manakin (*Pipra coronata*) populations from opposite banks of both the Napo and Amazon rivers have distinct genetic differences (Capparella 1985). The Amazon River largely isolates two similar species of antbirds. The dusky antbird (*Cercomacra tyrannina*) occurs north of the Amazon, while the similar blackish antbird (*C. nigrescens*) occurs south of the river. Both species occur together only along a section of the northern bank of the river, but here the blackish inhabits wet varzea forests, and the dusky favors second growth vegetation of the terre firme (Haffer 1985). Because of complex topography, especially in northern and western South America, geographic isolation has provided an ideal template for speciation. It is no accident that the Neotropics has a bird species richness of approximately 3300 (Haffer 1985), highest in the world. Diverse habitats ranging from lowland moist forests to alpine puna and paramo, plus tracts of habitat topographically isolated from similar tracts, provide ideal situations for promoting genetic divergence and speciation.

A perusal of any of the current guides to birds for areas in South America reveals almost extreme similarity among species within certain groups, evidence that speciation has probably been relatively recent. This is one reason why identification to species level is often so difficult. The antshrikes and antbirds, which are discussed more in chapter 6, illustrate well the fact that much of the species diversity of the tropics involves many species that look remarkably alike. Antbirds and antshrikes of the genera *Thamnophilus*, *Percnostola*, *Cercomacra*, and *Thamnomanes* are examples. All are basically little gray and brown birds of the rain forest understory. Though bird listers will not approve of this comment, I believe it fair to say that "you see one *Thamnophilus*, you've seen them all." Males of all species are sparrow-sized, slate gray birds (the intensity of the gray varies with species) with varying amounts of white spots on the wings and tail. Females are chestnut brown with or sometimes without white spots, depending on species. Such close morphological similarity among groups of species suggests a recent speciation (see below). Additional patterns of similarity among species groups are seen in the woodcreepers, hermit hummingbirds (see chapter 6), and many of the tyrant flycatcher genera.

J. V. Remsen, Jr. (1984) has recently described a pattern he termed "leapfrog" geographic variation among certain groups of Andean birds. He noticed a discontinuity in the appearance of three subspecies of the the mountain tanager *Hemispingus superciliaris*. A subspecies is a population not yet reproductively isolated from other populations within the species but different enough genetically that it does look distinct. The word *race* is essentially equivalent to subspecies. Subspecies formation is often, but not always, a precursor to speciation, and it is

Male antbird. Many species look
very similar, with various amounts of
white on wings and flanks.

Female antbird. Among the species,
females are, like males, very similar
in appearance.

often quite difficult to know whether or not different subspecies are interbreeding or are, in fact, reproductively
isolated (in which case they should be classified as separate species). The northern *H. superciliaris* subspecies in
Colombia and Ecuador is very yellowish, a pattern shared
with the subspecies in southern Peru and Bolivia. However, the subspecies of central Peru, which occurs *between*
the yellowish northern and southern subspecies, is dull
gray. The northern, central, and southern Andean subspecies are all closely related. Indeed, they are still considered members of the same species. Why should the central population diverge in color from the other two?
Remsen postulated that the complex topography of the
Andes has largely isolated the three subspecies and thus
reduced gene flow among them. He further postulated
that occasional random mutations, such as one for plumage color, could occur and not necessarily be selected for
or against. In short, subspecies could form and diverge
by chance, producing the odd "leapfrog" pattern of a
central population looking distinct from two bordering
populations, both of which look essentially alike. Remsen
surveyed other bird species with three or more distinct
subspecies and found that the leapfrog pattern was ap-

parent in about 20% of those he surveyed. He also examined Andean spinetails (genus *Cranioleuca*), small brownish birds with sharp pointed tails. He found that the crown color (top of the head) of one species had changed from white to predominantly buffy from 1936 to 1977, indicating a possible rapid evolutionary change. Many species of Neotropical birds exhibit complex patterns of subspeciation (Haffer and Fitzpatrick 1985).

The Ice Age and Tropical Refugia

The plants and animals of South America have a dynamic evolutionary history, tied to the equally dynamic geology of the continent (Raven and Axelrod 1975). Perhaps the greatest stimulus to speciation in the American tropics was provided by the Ice Age, a period known as the Pleistocene, which began about two million years ago. This sounds odd at first because most people naively believe that the equatorial tropics were stable and constant throughout the Ice Age, undisturbed by the giant glaciers bearing down upon northern temperate areas. Such was not the case, a point recognized by Thomas Belt who, in 1874, discussed possible effects of glaciation on the tropics. Jürgen Haffer (1969, 1974) and Beryl Simpson (Simpson and Haffer 1978) have brought together much evidence from geomorphology (the historical development of present landforms) and paleobotany (the study of past patterns of plant distribution) indicating that dramatic changes took place in Amazonia during the Ice Age.

Haffer and Simpson hypothesized that during glacial advance the Neotropics became cooler and drier. The ecological result of the climatic shift was that grassland savanna areas enlarged at the expense of forest tracts. Large continuous tracts of lowland rain forest were fragmented into varying-sized "forest islands" surrounded by "seas" of savanna. Because of the shrinking and fragmenting of forests, forest organisms became isolated from other populations in other forest areas. South

America became an "archipelago" of forest islands. Haffer and Simpson have postulated that nine major and numerous smaller forest islands were present in Amazonia during the Pleistocene and have termed these islands *refuges*, because they provided the last remaining suitable habitat for rain forest species. Because the refuges were geographically separated by wide expanses of grassland savanna, populations were sufficiently isolated from those in other refuges that conditions for speciation (and subspeciation) were ideal. Many groups thus went through periods of rapid speciation because there were repeated episodes of rain forest shrinkage and expansion. (During interglacial periods, forests expanded and secondary contact was established between newly speciated populations.) The importance of the refugia hypothesis is that it attributes a significant part of the high species richness of the tropics to the effects of *instability* rather than the older concept of the stable undisturbed tropics. Speciation was stimulated by geographic isolation brought about by the scattered refugia.

The refugia model has been examined in detail (see Prance 1982a), and evidence based on present distribution and diversity patterns does tend to add support. For example, Ghillean T. Prance (1982b) has closely examined woody plant diversity and concluded that twenty-six probable forest refuges existed for these plants. Warren G. Kinzey (1982) concluded that present primate distribution fits predictions of the refugia model and Haffer (1974), Haffer and Fitzpatrick (1985), and David L. Pearson (1977, 1982) have presented strong evidence from present bird distribution in support of the refugia model. On the other hand, not all present species distributions support the model. A complex of frog species of the genus *Leptodactylus* exhibits high species richness dating back farther than the Pleistocene. Most of the frogs speciated in the mid-Tertiary period, before the Ice Age began and refugias were created (Maxson and Heyer 1982; Heyer and Maxson 1982).

Though many speciation patterns fit the refugia model,

not all do and virtually all evidence in favor of the past existence and importance of refugia is circumstantial. It would be most helpful if pollen profiles from lakes or bogs could be studied to ascertain if, in fact, savanna and forest did oscillate in area, as refugia theory contends. There are few pollen profiles from South America, and these date back only 10,000 years, too recent to adequately test the refugia model (Connor 1986). Thus far no conclusive palynological evidence has been identified to unambiguously support or refute the refugia model. Edward F. Connor (1986) and others argue that present distribution patterns are open to other interpretations, and, thus, the refugia model for Neotropical diversity remains speculative.

ARE THE TROPICS IN EQUILIBRIUM?

Forest shrinkage and expansion during the Pleistocene may have helped shape present day patterns of Neotropical species richness, but to what degree are the tropics now stable? Are tropical ecosystems today in a state of equilibrium, as saturated with species as they can be? Some researchers believe they are. This view envisions the tropical rain forest to be maximally packed with specialized species, the vast majority of which have narrow ecological niches. This argument is really a restatement of Dobzhansky's hypothesis (see above). In recent years, however, evidence has been accumulating to suggest that rain forest ecosystems are not in equilibrium, though they are indeed packed with species. It is time now to again take up the question of why there are so many tree species in Neotropical forests.

Recent surveys in Amazonia support the contention that tree species richness is extremely high (Gentry 1988). Alwyn Gentry surveyed a series of 1-hectare forest plots in Peru and along the Venezuela-Brazil border. His most species-rich site was Yanomamö, Peru, where he recorded 580 individual trees of 283 species, all on but a single hectare! On this site, 63% of the tree species were represented by but a single individual. Gentry speculated that upper Amazonian forests may contain the most diverse floral and faunal assemblages on earth.

Why are tropical tree species so widely dispersed? Daniel Janzen (1970) hypothesised that a "seed shadow effect" occurs where seeds near a parent tree are subject to high rates of predation compared with seeds that somehow manage to be dispersed well away from the parent tree. Thus the surviving seedlings are only those located far from the parent, producing a wide dispersion pattern for each species. Evidence for this hypothesis remains unconvincing (see below).

It is also difficult to explain high tree species richness with the argument that each tree species has its own unique ecological niche (but see Ricklefs 1977). Suppose, however, that intermittent, moderate-scale disturbances such as hurricanes, lightning strikes, landslips, or insect plagues, occurring fairly frequently, have prevented competition among species from proceeding sufficiently far to result in extinction of "loser" species. For instance, if five tree species are in competition in an area of forest, given sufficient time, one species ought to predominate and outcompete the others. However, a disturbance, such as a hurricane, would perhaps annihilate all adult trees, thus removing the competitive edge of the "winning" species. The seeds and seedlings of all five species survive, however, and recolonize. Without the hurricane, a single species eventually prevails, and the site would be in equilibrium with one species, but, with the hurricane, the "game of competition" is restarted, and five species continue to occur on the site, not just one. This hypothesis is somewhat similar to the predator hypothesis (see above) except that natural disturbance is the "predator" in this case, and the "predator" is nonselective.

Joseph Connell (1978) believes that the frequency of disturbances is sufficiently high to maintain the tropics largely in a nonequilibrium state with competition not having time to reach its end point, namely, the extinction of some of the competitors. Thus, the tropics remain diverse. Connell argues that for groups such as trees, it is simply impossible to divide up the resources finely enough to permit equilibruim coexistence of dozens of species on a single site. After all, trees only compete for light, moisture, and basically the same array of minerals. Thus, the packing of species into a site is probably not due to narrow nonoverlapping niches. Rather, trees have wide overlapping niches, and species are in competition. Tree competition takes a long time, however, because life cycles are long. Periodic disturbance sets back the competition before it goes to completion. Thus, the tropics remain not only diverse but also in a nonequilibrium state.

Studies of coral reefs as well as tropical moist and dry forests provide evidence in support of Connell's disturbance-diversity hypothesis (Stoddart 1969; Connell 1978). Areas of coral reef in Belize that have not been struck by hurricanes have lower diversities of coral and other species than areas that have experienced disturbances. Coral species compete for space and sunlight (because some contain photosynthesizing algae), and, without disturbance, some coral species outcompete and exclude others. Disturbance, however, opens up the area and provides sites for many species to colonize. A major hurricane created highly patchy distributions on Jamaican coral reefs, supporting the notion that more mature reefs show high levels of species heterogeniety due to a past history of disturbance (Woodley et al. 1981).

Certain rain forest sites have been observed to have only one or a few species of canopy trees, in stark contrast to the usual high diversity observed. Connell notes that many ecologists have attributed the low diversity to uniquely poor soils. However, he argues that these low diversity sites are not always on poor soils and may, instead, be representative of the final outcome of long-term interspecific competition in the absence of disturbance. They are in equilibrium, and the equilibrium state is lower in diversity than the nonequilibrium successional state. On some of these low diversity sites, the seedling trees are of the same species as the trees of the canopy, indicating that equilibrium, a self-replacing ecosystem, has been reached.

Stephen P. Hubbell (1979) described the pattern of tree species distribution in a Costa Rican tropical dry forest and concluded that periodic disturbances were strongly affecting the sixty-one tree species present in the study area. Trees were randomly distributed and clumped, and rare species had low reproductive success. Hubbell rejected the seed shadow hypothesis and argued that periodic community disturbance was the factor most responsible for the distribution pattern. Hubbell (1980) has also found little support for the seed shadow hypothesis in moist tropical forests, arguing that they, too, are more subject to other factors.

Hubbell and Foster (1986c) suggest that combined effects of diffuse interspecific competition, periodic disturbance, and historical factors have resulted in the high species richness patterns evident throughout the Neotropics. This conclusion suggests that most tree species are ecological generalists, not specialists. Hubbell and Foster envision diffuse competition as being strong among species but resulting, not in the evolution of species with narrow niches, but rather in generalists occupying broad adaptive zones, each capable of rebounding from occasional disturbance.

> It is suggested that a common outcome of spatially and temporally uncertain competitors is likely to be a diffuse coevolution of generalist tree species within a few major life history guilds, rather than the pairwise coevolution of specialists in competitive equipose. As a result, tropical forests may accumulate tree species in part because forests of generalists may not be particularly resistant to invasion, and in part because there may be few forces besides drift that can lead to the systematic elimination of generalist tree species once established. (Hubbell and Foster 1986c)

In a study comparing several temperate and tropical forests, J. J. Armesto and two colleagues concluded that forests, whether in the tropics or temperate zone, that are subjected to frequent large scale disturbances such as landslides, fire, vulcanism, and hurricanes show a predominance of random distributions among tree species, a conclusion similar to Hubbell's (above). In forests rarely subjected to catastrophic disturbance (which are more common in the tropics), trees tend to be clumped, and the process of gap creation and colonization is of major importance. In a forest dominated by "gap-phase dynamics," there is a shifting mosaic of patches within the forest as gaps are created, colonized, and closed (Armesto et al. 1986).

James Karr and Kathryn E. Freemark (1983) present evidence for nonequilibrium in Panamanian rain forest bird communities. They found that birds are highly sensitive to both moisture gradients and gradients of vegetation structure. Differences in microclimate from site to site were of major influence in determining which bird species occurred. Since microclimates change regularly so does the bird community at any given site. Karr and Freemark emphasized that though the bird communities vary with microclimate changes, and are thus not in equilibrium, the changes in the bird communities are not random. Bird species respond individually in predictable ways to microclimate and vegetation change. What is not predictable is what the composition of the entire community will be. This is what fluctuates with time due to changes in vegetation and microclimate.

The Neotropics are dynamic. Past climate shifts may have been a major factor causing rain forest shrinkage and stimulating episodes of speciation. Today, frequent storms, windthrow, and other disturbances create gaps (see chapter 3), reinitiate succession, and preserve high diversity. With disturbance and nonequilibrium added to an already complex picture of tropical diversity discussed earlier in the chapter, it becomes apparent that many factors must combine to produce and preserve these most diverse of ecosystems. Factors affecting the diversity of one taxon may not be those affecting another, at least not to the same extent. It is probably quite hopeless to search for a single simplistic answer to the question, Why are there so many species in the tropics?

COEVOLUTION

Species affect each other in intricate ways. Herbivores are dependent on plants as food, but plants are spiny and often toxic, an apparent defense against herbivores (see the next chapter). On the other hand, herbivores help recycle nutrients by their waste products, serve as pollinators, and aid greatly in seed dispersal. The relationship between herbivores and plants is anything but simple. Predators are adapted to capture prey, but prey are adapted to avoid capture. The ecological "cat and mouse game" between predators and prey has given rise to elegant adaptations including both cryptic and warning coloration. The "ecological web" woven by the myriads of interactions among species within an ecosystem has an important historical aspect. As mentioned earlier, biotic selection pressures, the influences of predators on prey, of herbivores on plants, parasites on hosts, etc., are important in determining the evolutionary directions taken by natural selection in the tropics. When one species has a trait that acts as a selection pressure on another, and the second species in turn possesses a trait that acts as a counter selection pressure back upon the first, the evolutionary fates of both species become locked together. This situation is called *coevolution* (Ehrlich and Raven 1964; Janzen 1980b; Futuyma and Slatkin 1983).

Pollination

One of the most obvious examples of coevolution is pollination. Most flowering plants are totally dependent upon insects, birds, and bats for survival. Animals seek out flowers and obtain nectar and, in some cases, pollen as a food. Animals, by their travels from plant to plant, disperse pollen, making cross pollination efficient and insuring reproduction of the plants. This relationship is termed *mutualistic* because both plant and pollinating animal benefit. Charles Darwin, in *On the Origin of Species*, wrote about the coevolved relationship between bees and the clover they pollinate: "The tubes of the corollas of the common red and incarnate clovers (*Trifolium pratense* and *incarnatum*) do not on a hasty glance appear to differ in length; yet the hive-bee can easily suck the nectar out of the incarnate clover, but not out of the common red clover, which is visited by humble-bees alone." Darwin discussed pollination and coevolution further in a monograph *On the various contrivances by which British and foreign orchids are fertilised by insects, and on the good effects of intercrossing*, published in 1862. He noted how closely the anatomies of some flowers and pollinating insects coincide. Certain orchids even resemble the backsides of female bees, and males, attempting to copulate with them, actually pollinate them.

In the tropics, animal pollination is widespread. Rain forests are sufficiently dense that wind pollination would tend to be ineffective, except perhaps for emergent trees. It is not surprising that grasses, sedges, pines, and other open area savanna species are the only tropical plant groups dominated by wind pollination. Vertebrates and insects are both major pollinators in the Neotropics. Hummingbirds and bats, for instance, are specialized to feed on many flower species. Hummingbird pollinated flowers tend to have rather long tubes and are red, orange, or yellow. Bat-pollinated flowers are often white (easy to locate in the dark) and frequently have a musky odor, an attractant to the bats. Many flowers are pollinated by a variety of vertebrates and insects. In Trinidad, I watched seven hummingbird species feed on vervain, which was also being visited by various bees and assorted flies. Vertebrate pollinators are advantageous to plants in that they tend to fly long distances. Such behavior helps insure effective cross-pollination between widely separated plants (Janzen 1975). Male euglossine bees are also long distance fliers, and they pollinate certain orchids that are widely separated. Compounds in the orchid flower that are absorbed by the male bees contribute to the longevity of the insects. Male euglossine bees can live up to six months, a long life for a bee (Janzen 1971b).

An elaborate coevolution has occurred between figs and fig wasps (Janzen 1979; Wiebes 1979). Various species of figs (*Ficus* spp.) all produce a bulbous green flower that, to the casual eye, appears to be not a flower at all but a fruit. The flower is called a synconium and is really a cluster of enclosed flowers within a gourdlike covering. Externally, the fig flower reveals nothing of its unusual inner structure. Internally, it is a dense carpet of tiny flowers, some male, some female, and some sterile (called gall flowers). Though male and female flowers exist side by side, males cannot pollinate females in the same synconium because female flowers mature earlier than males. Since the synconium is utterly enclosed, it would appear impossible for fig pollination to occur without the assistance of an animal pollinator, and a mighty skilled one at that. Enter (literally!) the tiny fig wasps (family Agaonidae).

Sterile gall flowers contain fig wasp eggs laid by females of the previous generation who died after depositing their eggs. Males hatch first and burrow into gall flowers to inseminate the still unhatched females. Each male commonly inseminates several females. Females hatch, already pregnant with the next wasp generation. Newly hatched females wander over the stamens of male flowers, which have reached maturity precisely when the females hatched. Laden with pollen, the female wasps either make an exit tunnel from the flower or, in some species, the flower opens enough to permit females to exit. Females have but a day or so to locate another flowering fig and tunnel into a flower. The female is often physically damaged from burrowing into the flower but, once inside, she has only to locate a gall flower in which to deposit her eggs. After that, she dies. In her search for a gall flower, the female passes over fertile female flowers and deposits the pollen she brought from the flower of her birth.

Parasitic fig wasps also inhabit synconiums. These insects use the fig as a food source but do not pollinate it. Parasitic males roam about in the utterly dark interior of the synconium using their huge jaws to dismember other parasitic males as they all search for females with which to mate. William Hamilton (1979), who spent a great deal of time looking into fig synconiums, wrote, ". . . a situation that can only be likened in human terms to a darkened room full of jostling people among whom, or else lurking in cupboards and recesses which open on all sides, are a dozen or so maniacal homicides armed with knives."

It is apparent that everything about the reproductive biologies of figs and fig wasps is interrelated. Such an intricate level of interaction indicates a long evolutionary association between plants and insects. This is coevolution.

Chiropterophily

Pollination of flowers by bats is common in the tropics, with over five hundred plant species wholly or partly dependent on bats as pollinators (Heithus et al. 1974). Plants adapted to host bats are termed *chiropterophilous,* meaning "bat-loving" (bats are in the mammalian order Chiroptera). Donna J. Howell (1976) has reviewed coevolution between bat-plants and plant-bats, a relationship revealing that coevolution has occurred at the physiological as well as anatomical levels in both bats and plants.

As already noted above, plants with bats as primary pollinators tend to have large white flowers, often with a musky "batlike" odor. These flowers, of course, open at night when bats are active. Flowers are often shaped like a deep vase, though some are flat and brushy, loading the bat's face with pollen as it laps up nectar. Many bat flowers are cauliflorous, growing directly from the tree trunks. Flowers may hang from long, whiplike branches (flagelliflory) or hang downward as streamers (penduliflory), a condition common in many vines. Cauliflory, flagelliflory, and penuliflory all have in common the fact that the flowers are positioned in such a manner that they are easily accessible to bats, which are generally large and must hover in the open as they feed. Working in Guanacaste Province, Costa Rica, E. Raymond Heithus and two colleagues (1974) noted that two bat species feeding on flowers of *Bauhinia pauletia* exhibited differing feeding behaviors. *Phyllostomus discolor* visited high flowers, grasping the branch beneath the flower and pulling it down. *Glossophaga soricina* visited both high and low flowers and hovered as it fed. Both bat species aided in cross-pollinating *Bauhinia*.

Nectar-feeding bats tend to have large eyes and relatively good vision. They visually locate white flowers that show up well in the dark. The sonar of nectar-feeders is often quite reduced, but the olfactory sense is well developed. Sight and scent, not sound, are how the animals find their sugary dinners. They also have long muzzles and weak teeth, both advantageous in probing deeply into flowers. Finally, they have long tongues covered with fleshy bristles that can extend well into the flower, and their neck hairs project forward, acting as a "pollen scoop." When nectar-feeding bats feed, they pick up a great deal of pollen.

The pollen from bat plants tends to be significantly higher in protein than in nonbat pollinated plants, and bats ingest pollen as well as sugary nectar. Pollen contains the amino acids proline and tyrosine, useless to the plant but important to the bats. Proline is necessary to make connective tissue such as is used in wing and tail membranes, and tyrosine is essential to milk production. Bats are not the only animals to eat and derive important protein from pollen. Heliconius butterflies (next chapter) collect pollen on a small brush near their mouthparts, ingesting it after dissolving it with nectar.

Once ingested, nectar helps dissolve the tough pollen coat, but bats aid this process because their stomachs secrete extraordinarily large amounts of hydrochloric acid. These bats also drink their own urine, which helps dissolve the pollen coat liberating the essential protein. Nectar-feeding bats and bat pollinated plants, like the figs and fig wasps, are locked together in a mutualistic relationship in which each party is essential to the other.

Ants and Ant Plants

Many tropical plant species possess nectar-secreting glands apart from flowers. Found on leaf blades, leaf petioles, stems, or other locations on the plant, these glands manufacture various energy rich sugary compounds as well as certain amino acids. Termed *extrafloral nectaries*, these odd sugar-producing bodies initally puzzled botanists, who could identify no obvious use for them. It was soon observed, however, that many of the plants with extrafloral nectaries are liberally populated by various ant species, most of which are aggressive. Perhaps the ants, by their aggressiveness, somehow protect the plants, which repay the ants in nectar. This rather startling hypothesis was termed the *protectionist hypothesis*, and it envisioned the relationship between the plants and ants as mutualistic. The alternative idea, called the *exploitationist hypothesis*, argued that the ants merely fed on the sugary nectaries but provided no actual service to the plants. Barbara L. Bentley (1977), quotes a 1910 description of the exploitationist hypothesis as believing that the plants "have no more use of their ants than dogs do their fleas."

Among the most common plants with extrafloral nectaries are the cecropias (chapter 2). Cecropias have nectaries, termed *Mullerian bodies*, at the base of the leaf petiole, where the leaf attaches to the stem. Ants of the genus *Azteca* live in the hollow pith of the stem and feed on the Mullerian bodies. I have encountered these ants frequently, and they are not nice ants. Cutting cecropias for use as net poles (to catch and band birds), I have been attacked vigorously by Aztec ants, and, thus, I have little sympathy for the exploitationist hypothesis. The ants of a cecropia seem both highly pugnacious and quite protective of their tree. If I were a cecropia, I'd want some of these ants living on me.

Daniel Janzen (1966) settled the controversy at least for one ant species. Janzen studied *Pseudomyrmex ferruginea*, which occurs on five species of *Acacia* tree. Commonly called the bull's horn acacia, the tree has pairs of large hollow thorns on its stem that serve as homes for the ants. A single queen ant burrows into a thorn in a sapling acacia to begin a colony that can increase to as many as 12,000 ants by the time the tree is mature. Janzen noted that by the time the fast growing tree was seven months old, 150 worker ants were "patrolling" the stem. The acacia ants attack any insects that land on or climb on the tree, including beetles, bugs, caterpillars, and other ants. Ants also clip plants that begin to grow nearby or overtop and shade the acacia (thus taking its sunlight), and they attack cattle or people if they brush against the tree. Ants become very active, swarming out of the thorns and over the foliage even if exposed to the odor of cattle or people. I was attacked on the neck by a single acacia ant that presumably dropped on me as I was lecturing in the field. The formic acid irritated me for over a day, and I can well imagine the discomfort that would have occurred if I had been stung by many of these ants.

Why do ants live on acacias? They obtain shelter within the thorns, but they also obtain nutrition from two kinds of extrafloral nectaries. One type is termed *Beltian bodies*, which are small orange globules growing from the tips of the leaflets of the compound leaves, and the other type is called foliar nectaries, located on the petioles.

Janzen performed a field experiment that discriminated between the protectionist and exploitationist hypotheses. He treated some acacias with the insecticide parathion, and he also clipped thorns to remove all ants from the treated trees. The antless trees did not survive nearly as well as control trees, which were permitted to keep their ants. Janzen estimated that antless acacias were not likely to survive beyond one year, either falling prey to herbivores or being overtopped by other competing species of plants. Ants are needed to attack herbivores and clip other plants. An antless acacia is doomed. Janzen concluded that the ants and acacias are *obligate symbionts*, depending entirely upon each other. The protectionist hypothesis is correct, and an impressive mutualism has coevolved between acacias and *Pseudomyrmex*.

Extrafloral nectaries occur also on temperate zone plants (Keeler 1980) but seem more abundantly represented among tropical species (Oliveira and Leitao-Filho 1987). In a survey of riparian and dry forests in Costa Rica, plants with extrafloral nectaries ranged in percentage cover from 10 to 80% (Bentley 1976). In the Brazilian cerrado, cover by woody plants with extrafloral nectaries ranged from 7.6 to 20.3% (Oliveira and Leitao-Filho 1987). Many plants with extrafloral nectaries house ants, but the degree to which they protect their hosts varies (Bentley 1976, 1977).

Fungus Gardens and Leafcutting Ants

No one can visit the tropics without encountering leafcutting ants. Throughout rain forests, successional fields, and savannas, well-worn narrow trails are traversed by legions of ants of the genus *Atta* as they travel to and from their underground cities, bringing freshly clipped leaves. Their trails take them up into trees, shrubs, and vines where they neatly clip off pieces of leaves that they carry back to their colony. Living in underground colonies of up to 5 million individuals, worker ants may be small (minimas) or medium-sized (medias). Soldiers are large (maximas) and well armed with formidable pincer jaws. *Atta* cities are underground, but the mounds of soil that mark their entrances can spread widely in an area of forest. The leafcutters make no secret of their presence. There are approximately two hundred leafcutter species, and, though most abundant in the tropics, leafcutters also occur in warm temperate and subtropical grasslands. One enterprising species even occurs as far north as the New Jersey pine barrens (Wilson 1971).

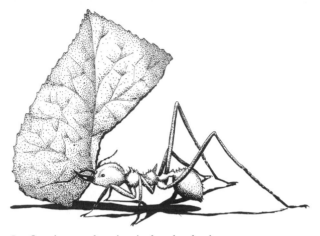

Leafcutting ant bearing its burden back to
the colony's nest.

The impact of leafcutting ants may prove enormous.
On Barro Colorado Island in Panama, leafcutter ants
have been estimated to consume 0.3 tons of foliage/hec-
tare/year, equal to the combined effects of all vertebrates
in the forest (Leigh and Windsor 1982). Larry L. Rockwood (1976) studied the selectivity of leafcutters in Guanacaste, Costa Rica. Examining two species of *Atta*, Rockwood found that one species clipped mature leaves from
only 31.4% of the plant species available. The other species used leaves from only 22% of the available plant species, indicating a strong selectivity of both leafcutter species. Leafcutters don't clip just any plants but only certain
species. The commonness or rareness of a plant species
has no correlation with *Atta* preference. They often will
travel far from their nest to seek out a certain plant.
Rockwood hypothesized that internal plant chemistry
strongly influences *Atta* diet, a suggestion borne out by
subsequent research (Hubbell et al. 1984). Ants seem to
concentrate on plants with minimal amounts of defense
compounds in their leaves (see chapter 5).

Although leafcutting ants taste the sap of the leaves

they cut, they do not consume any leaves but merely clip and carry leaf fragments to their colonies. There they use the leaves to make media to culture a fungus. This odd fungus, which is never found free-living outside of *Atta* colonies, is the ants' only food, hence their other name, fungus garden ants. Leaves brought to the colony are clipped into small pieces and chewed into a soft pulp. Before placing the pulpy mass on the fungus bed, the ant holds it to its abdomen and defecates a fecal droplet of liquid on it. The chewed leaf is then added to the fungus growing bed and small fungal tufts are placed atop it. Other ants sometimes add their fecal droplets to the newly established culture.

Worker ants collecting leaves avoid those that contain chemicals potentially toxic to the fungus (Hubbell et al. 1984). One tree, *Hymenaea courbaril*, a legume, has been shown to contain terpenoid (see next chapter) that is antifungal (Hubbell et al. 1983). *Atta* ants avoid clipping leaves from this species. The tree has evolved a protection from *Atta*, not by poisoning the ant, but by poisoning its fungus!

The relationship between ants and fungus is unique. The ants only culture a few fungal species, all of which are members of one group, the Basidiomycetes. The fungi are always in pure culture, protected from contamination from other fungal species by constant "weeding" by ants. As mentioned earlier, the few fungal species cultured by *Atta* species are never found growing outside of the ant colonies. Both ants and fungi are totally dependent on each other. They represent a complete obligatory mutualism (Weber 1972). Ant and fungus are inextricably linked by evolution: when a queen *Atta* founds a new colony she takes some of the precious fungus with her inside her mouth. Fungus and ants disperse together.

Michael M. Martin (1970) has made detailed studies of the fungus-ant relationship at the biochemical level, and his work has revealed the specific role that ants play in culturing the fungus. The ants clean the leaves as they chew them to make the culture bed pure. Ant rectal fluid contains ammonia, allantoic acid, the enzyme allantoin, and all twenty-one common amino acids. These compounds are all low molecular weight nitrogen sources, and they are the key ingredients in making the culture optimal for the fungus. The fungus lacks certain enzymes that break down large proteins (all of which are made up of chains of amino acids). Thus it depends totally on the ant rectal fluid to supply its amino acids. Experiments attempting to grow the fungus in a rich protein medium failed. It can only grow in a medium of small polypeptides and amino acids. Ants also supply enzymes necessary to aid in breaking down protein chains.

Martin summarized the functions of the ants as (1) dispersal and (2) planting of the fungus, (3) tending the fungus to protect it from competing species, (4) supplying nitrogen in the form of amino acids, and (5) supplying enzymes to help generate additional nitrogen from the plant medium. The ants are the expert gardeners of the insect world, and their labors pay off handsomely. The fungus can digest cellulose, an energy rich compound that ants cannot digest. By eating the fungi, ants can tap into the immense abundance of energy in rain forest leaves. Fungus garden ants owe their remarkable abundance to their unique evolutionary association with fungi.

How such an intricate evolutionary relationship began is difficult to know, but ant species do exist that culture other fungi on beds of chewed leaves and flowers as well as on insect carcasses. These relationships are not nearly so interdependent as the *Atta*-fungus relationship, but they do indicate that intermediate complexities are possible during the evolutionary process linking insect and fungus together. It is probable that the *Atta*-fungus relationship began through simple predation by ants on fungus. What began as predation evolved over the millennia into mutualism.

THE IMPORTANCE OF FRUIT IN THE TROPICS

Fruit is both abundant and constantly available throughout the year in the tropics, making it an important resource, especially for birds. In the temperate zone, fruit is a seasonal resource, occurring from midsummer through autumn. Many birds migrating to winter in the tropics feed heavily on fruit in the fall, but, because fruit is ephemeral in the temperate zone, no bird families have specialized as *frugivores* (E. W. Stiles 1980, 1984). In the tropics, entire families of birds such as the manakins, cotingas, toucans, parrots, and tanagers depend heavily on fruit, and some species are exclusively frugivores (Snow 1976; Moermond and Denslow 1985). In addition, mammals ranging from bats to agoutis to howler monkeys utilize fruit as a major component of their diets.

The function of a fruit is to advertise itself to some sort of animal so that it will be eaten. The seed(s) then passes through the alimentary system of the animal (or is regurgitated), and, because the animal is mobile, seeds are deposited away from the parent plant. Fruits help insure seed dispersal. The animal derives nutrition from the fruit but also disperses seeds; thus, the relationship between animal and plant is mutualistic. It is to the plant's ultimate advantage to invest energy in making fruit and to the animal's immediate advantage to eat the fruit. Some parrot and pigeon species, however, digest the seed as well as the pulp of the fruit (or else they injure the seed, and it does not germinate). These "seed predators" are not useful as dispersers and must be considered parasites rather than mutualists.

Some plants do not "invest" in expensive fruits to attract animals as seed dispersers. Some rain forest canopy trees, vines, and epiphytes utilize wind dispersal of seeds, though many other species are animal-dispersed. Daniel Janzen (1983a) points out that wind dispersal is most common at the canopy level or in deciduous forests, where leaf drop can help facilitate wind movement of seeds. Janzen cites a study in Costa Rica conducted by H. G. Baker and G. W. Frankie, who surveyed 105 tree species in a deciduous forest and found that 31% were clearly wind-dispersed. In dense interior rain forests animals are far more important than wind for seed dispersal.

Ornithologists David W. Snow (1976) and Eugene Morton (1973) have considered the evolutionary consequences to birds of a diet mainly of fruit. Fruit is generally a *patchy resource*, meaning that it may be abundant on a given tree (for instance, a fruiting fig tree), but trees laden with mature edible fruits are widely spaced in the rain forest. Snow notes that such a resource distribution selects for social behavior rather than individual territoriality. Flocks of fruit eaters can locate fruiting trees more efficiently than solitary birds, and there is no disadvantage to being in a flock once the fruit is located since there is usually more than enough fruit for each individual. It is extremely unusual, for instance, to see a single parrot or a single toucan. They always come in flocks. Snow studied the purple-throated fruit crow (*Querula purpurata*) in Guyana, a species that feeds on both insects and fruits. These birds live in small communal groups of three or four individuals that roam the forest together in search of preferred species of fruiting trees. Within the social group there is virtually no aggression, and all members of the group cooperate in feeding the nestling bird (they have clutches of only one). The nest is in the open and is vigorously defended by the entire group (Snow 1971). Fruit eating mammals also tend toward sociality. Pacas, coatimundis, and pecarries roam about in bands.

Snow points out that frugivorous birds have much "free time," since fruits are generally easy to locate and require virtually no "capturing" time and effort. The male bearded bellbird, a frugivore that I will discuss in more detail in chapter 6, spends an average of 87% of its time calling females to mate. Another frugivore, the male white-bearded manakin, spends 90% of its day courting females! Snow adds that frugivorous birds are highly abundant compared to insectivorous species, again, because fruit is abundant and readily available. Insects, on the other hand, are widely dispersed, often difficult to find and capture, and represent far less overall biomass. In one area of rain forest in Trinidad, Snow netted 471 golden-headed manakins and 246 white-bearded manakins, a total of 717 birds. In this same area he caught eleven species of tyrant flycatchers, but their combined total did not equal that of the two frugivorous manakins. Snow's wife, Barbara K. Snow, made a detailed study (1979) of the ochre-bellied flycatcher (*Pipromorpha oleaginea*) on Trinidad (page 145). As mentioned earlier, this species is undergoing an evolutionary diet shift. Though it is a tyrant flycatcher, it feeds almost exclusively on fruit. The Snows found the ochre-bellied flycatcher to be the most abundant forest flycatcher in Trinidad and attributed its numerical success to its diet of fruit. Fruit will support more of a given species than will a diet of insect food. In Belize, my colleague William E. Davis, Jr. and I have found the ochre-bellied flycatcher to be by far the most abundant of the twenty-three flycatcher species we encountered. We netted 102 ochre-bellies compared with 14 sulfur-rumped flycatchers (*Myiobius sulphureipygius*), the next most frequently netted flycatcher species.

A diet of fruit does pose potential problems, however. Interspecific competition may occur due to the localized nature of the resource (Howe and Estabrook 1977). Nutritional balance may be lacking. Seeds are usually undigestible, and fruits tend to be highly watery and contain little protein compared with carbohydrate (Moermond and Denslow 1985). Small birds tend to eat small, carbohydrate-rich fruit, and many diversify their diets to include arthropods. Large frugivores, such as toucans, eat many different sized fruits including those rich in oil and fat (Moermond and Denslow 1985). Many of these species also suppliment their diets to include animal food.

Who Gathers at the Fruit Tree?

Fruiting trees attract many species. One of the best ways to see both birds and mammals is to locate a tree laden with fruit and watch what comes by to feed. On January 14, 1982, I observed the following seventeen species of birds at a single fruiting fig tree in Blue Creek Village, Belize: masked, yellow-winged, and blue-gray tanagers; masked tityra; clay-colored robin; black-and-white warbler; streak-headed woodcreeper; yellow-throated euphonia; red-legged honeycreeper; northern and orchard orioles; buff-throated saltator; collared aracari; aztec parakeet; lovely cotinga; social flycatcher; and black-cheeked woodpecker. Not all of these species fed directly upon the fruits because insects are also attracted to fruit. Both frugivores and insectivores benefit from a visit to a fruit tree. Charles Leck (1969) observed sixteen species of birds ranging over eleven families on a single tree species (*Trichilia cuneata*) in Costa Rica. Leck noted both intra and interspecific aggression among various birds, and, of the sixteen species, eleven were observed directly feeding on fruit.

Henry F. Howe (1977) studied the fruit dispersal of *Casearia corymbosa* in Costa Rica and noted twenty-one species feeding on the fruits, none of which really contributed to seed dispersal. These species failed to move significantly away from the fruiting tree. Only one bird species, the masked tityra, *Tityra semifasciata*, was deemed by Howe to be an efficient seed disperser. This robin-sized black and white bird fed heavily on *Casearia* fruits

Masked tityra, a seed-disperser.

and regurgitated viable seeds at considerable distance from the parent tree. Howe noted that the tityra is common and has a high feeding rate, characteristics enhancing its efficiency as a seed disperser. Howe (1982) also studied the tree *Tetragastris panamensis* on Barro Colorado Island and found an assemblage of twenty-three birds and mammals that fed on the fruits, which were produced in "spectacular displays of superabundant but (nutritionally) mediocre fruit. . . ." Howe noted that actual seed dispersal of this species was less effective than another species, *Virola surinamensis*, which produced fewer but nutritionally better fruits. Fewer bird species fed on *Virola*, but those that did regurgitated seeds one at a time, well away from the plant where they were ingested.

Studies in Costa Rica by Douglas J. Levey, Timothy C. Moermond, and Julie S. Denslow (Levey, Moermond, and Denslow 1984; Levey 1985; Moermond and Denslow 1985) indicated that two patterns are evident in methods by which birds take fruit. Calling them *mashers* versus *gulpers*, the researchers noted that some birds mash up the fruit, dropping seeds, while others gulp the fruit, ingesting it whole, and either regurgitate or defecate seeds. Mashers are finches and tanagers, and gulpers are toucans, trogons, and manakins (see chapter 6). Mashers are more sensitive to taste than gulpers. Levey found a distinct preference among masher species for 10–12% sugar solutions as opposed to lower concentrations. Gulpers, which swallow fruit whole, are taste insensitive. Gulpers have wide wings aiding them in hovering at the fruit tree, and they have wide flat bills useful in plucking and swallowing the fruits.

Precise measurements of fruit volume and seed dispersal were reported by Pedro Jordano (1983), who observed a single fig tree in a lowland deciduous forest in Costa Rica. Jordano estimated the total fig crop to be approximately 100,000, all of which were taken within five days. During the first three days alone, 95,000 were devoured! Birds were the principal feeders, eating 20,828 figs each day, which Jordano calculated to be about 65% of the daily loss. Mammals and fruits falling to the ground were the other source of loss to the tree. Parrots, which are seed predators, accounted for just over 50% of the daily total of figs. The most efficient seed dispersers, orioles, tanagers, trogons, and certain flycatchers, took only about 4600 figs per day. Considering seed number, Jordano calculated that approximately 4,420,000 seeds were destroyed each day, mostly from predation by wasps and other invertebrates. Parrots were estimated to account for 36% of the seed loss. Jordano estimated that only 6.3% of the seeds lost from the tree each day were actually dispersed and undamaged by birds. Jordano's detailed study indicates the high cost of seed dispersal.

Nathaniel T. Wheelwright (1984) and three colleagues conducted a comprehensive survey of fruiting trees and fruit-eating birds at Monteverde, a lower montane forest in Costa Rica. They found that 171 plant species bore fruit that was fed upon by 70 bird species. Some birds depended heavily on fruit, others casually. Among the birds were 3 woodpecker species, 9 tyrant flycatchers, 8 thrushes, 8 tanagers, and 9 finches, as well as toucans, pigeons, cotingas, and manakins. Though some birds were observed to feed on fruit only rarely, it was clear from

The head of a mealy parrot, a seed-predator.

this study that fruit represents an important resource for a very large component of the avian community.

Wheelwright (1985) also reported that birds are selective about the size of the fruits they eat. He observed species such as toucanets pluck fruit, juggle it in their bills, and then reject it by dropping it. Large fruits are particularly at risk of rejection, and Wheelwright found many large fruits scarred by bill marks beneath the tree. Wheelwright hypothesized that plants are under strong selection pressure to produce small to medium-sized fruits, as larger ones are rejected by most bird species except those with the widest gapes. Since wide-gape species were observed to feed just as readily on small as on large fruits, plants that produce large fruits must depend upon there being many large-gape species to insure seed dispersal. Large fruits do permit more energy to be stored in the seeds, an advantage once dispersal and germination have occurred.

The Oilbird—A Unique Frugivore

My choice of the most unique frugivore is the bizarre oilbird, *Steatornis caripensis*, often called the *guacharo*. Only one species exists in this odd family (Steatornithidae), which ranges from Trinidad and northern South America to Bolivia (Hilty and Brown 1986). The oilbird is a large nocturnal bird, its body measuring 18 inches in length and its wingspread nearly 3 feet. It is a soft brown color with black barring and scattered white spots, and it has a large hooked bill and wide, staring eyes. Oilbirds are fascinating enough as individuals, but they come in groups. Colonies live in caves, the birds venturing out only at night to feed on the fruits of palms, laurels, and incense, often obtained only after flying up to 30 miles from the cave (Snow 1961, 1962b, 1976). Fruits are taken on the wing as the birds hover at the trees, picking off their dinners with their sharp hooked beaks. Oilbirds probably locate palms by their distinct silhouettes, but

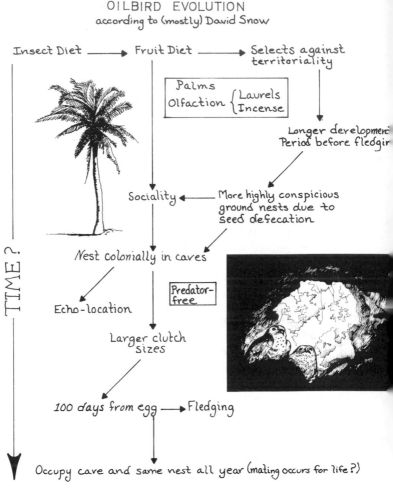

OILBIRD EVOLUTION
according to (mostly) David Snow

Insect Diet ⟶ Fruit Diet ⟶ Selects against territoriality

Palms
Olfaction { Laurels
Incense

Longer development
Period before fledging

Sociality ⟵ More highly conspicuous ground nests due to seed defecation

Nest colonially in caves

Predator-free

Echo-location

Larger clutch sizes

100 days from egg ⟶ Fledging

Occupy cave and same nest all year (mating occurs for life?)

TIME?.

A schema of the possible events in the evolution of the oilbird.
See text for details.

they home in on laurels and incense through odor. Fruits from these trees are highly aromatic.

Enter an oilbird cave, as I have in Trinidad's Arima Valley, and you have quite an experience in store. The birds explode in a cacophony of sound somewhere between a growl and a scream. In the dark, dank cave the flapping wings of the disturbed, screaming host conjures up thoughts of tropical demons awakened. Soon the birds flutter restlessly back to their nesting ledges, snarling as they resettle. Among the din you hear some odd clicklike noises. These vocalizations are one of the reasons why oilbirds are unique. They are the only birds capable of echolocation, the same technique by which bats find their way in the dark. The clicks are their sonar signals, sent out to bounce off the dark cave walls and direct the birds' flight. Oilbird echolocation lacks the sophistication found in bats, but it serves well to keep the birds from crashing against the cave walls as they fly in total darkness.

Oilbirds are closely related to nightjars of the order Caprimulgiformes. These nocturnal birds, of which the whip-poor-will (*Caprimulgus vociferous*) is a common example, differ from oilbirds in that all are insectivorous, none are colonial, nor do any live in caves. How did the oilbird evolve frugivory, sociality, and its cave-dwelling habit? David Snow (1976) has suggested a scenario for oilbird evolution that begins with a shift in diet from insects to fruit. Snow hypothesizes that oilbirds were originally "normal" nightjars feeding on insects. However, large fruits offer a potentially exploitable resource, especially to a night bird since there are few large bats and no other nocturnal birds to offer competition. Oilbirds became oilbirds when their ancestors shifted to a new diet of fruit.

Frugivory selected for several adaptations. First, the birds' olfactory sense became enhanced as that provided an advantage in locating aromatic fruits. Second, social behavior began to evolve because fruits are a patchy resource and birds would tend to come together at fruiting trees. More important, however, is that a diet of fruit, though rich in calories, is nutritionally unbalanced, and, consequently, the nesting time is quite prolonged. Incubation of each egg lasts just over one month. Once oilbirds hatch, they fatten up immensely in the nest due to a buildup of fat from the oily fruits, though they take a very long time to acquire sufficient and proper protein to grow bones, nerve, and muscle. By the time a juvenile is two months after hatching, it has still not left its nest but may weigh 1.5 times what either of its parents weigh! The name *oilbird* refers to the fact that juveniles put on so much temporary fat that they can be boiled down for the oil. Indians also occasionally use them for torches since they burn so well! The total time it takes an oilbird to go from newly hatched egg to fledging and independence is nearly one hundred days. For a full clutch of four eggs to develop requires approximately six months. Such an extended development time makes it very risky to nest on the forest floor, the traditional nightjar nesting site. Also, Snow notes that the defecated seeds that would surround a ground nest would serve to bring attention to it. Cave dwelling offers much more protection for the nest, but caves are also very patchy resources. Again, sociality is selected for, and oilbirds became colonial cave dwellers.

Cave dwelling selected for the development of echolocation and also for a larger clutch size. Most tropical birds lay very few eggs in a given nest, often only one (like the purple-throated fruit crow mentioned above). Predator pressure is likely to be a major reason for the small clutch sizes, since nests can be more secretive if there are fewer mouths to feed. Given the safety of caves, however, oilbird nests are not under severe predator risk and oilbird clutch size is normally four eggs. The nest is built up using droppings and located on a cave ledge. Thin, yellowish, light-starved seedlings sprout from defecated seeds around the nest. Birds are thought to pair for life.

The scenario given above for oilbird evolution may or may not be true. David Snow has provided an evolutionary scenario accounting for all major oilbird adaptations,

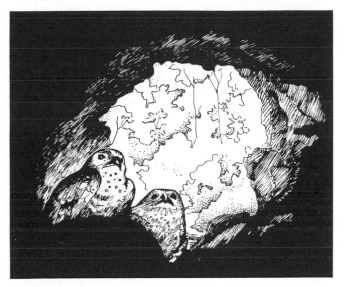

An oilbird pair.

but since it is not possible to go back in time, it is not possible to rigorously test Snow's hypothesis. The oilbird remains one of the most fascinating of tropical species, and visiting an oilbird cave is one of the best reasons for making a trip to the tropics.

CHAPTER 5

The Tropical Pharmacy: Plant Drugs and Their Consequences

TROPICAL rain forests are green. Myriads of leaves, large, small, simple, and compound, adorn trees and vines from ground to canopy. Careful examination of leaves reveals that most show little if any damage from insects or other herbivores.

At least in theory, tropical rain forest leaves are subjected to potential damage by insects and pathogens year-round. There is no cold winter season when arthropods are mostly inactive. Yet leaves remain largely intact. In one Amazon rain forest, biomass of living vegetation was estimated at approximately 900 metric tons per hectare. The mass of animals, all mammals, birds, reptiles, insects, and other creatures totaled a mere 0.2 metric tons per hectare, thus the biomass of animals was only 0.0002% of the plant biomass! Clearly the plants have the animals under control. Of the total number of animals in this Amazonian forest, only 7% ate living plant material such as leaves and stems; 19% (mostly termites) depended on eating living or dead wood; and fully 50% ate only dead vegetation. The remainder were carnivores (Fittkau and Klinge 1973). In general, herbivores in the tropics take only a small percentage of what is potentially available to them (Coley et al. 1985). How do tropical leaves protect themselves against the herbivorous hordes as well as against invasive pathogenic bacteria and fungi?

PLANT DEFENSE COMPOUNDS

The answer is drugs. Leaves of both tropical and temperate zone plants are laced with chemicals, some of

Tropical leaves are often relatively
untouched by insect damage.
Defense compounds help discourage
herbivores. See text for details.

which stir human emotions ranging from fear to desire.
Daniel Janzen (1975) puts it well: "The world is not col-
oured green to the herbivore's eyes, but rather is painted
morphine, L-DOPA, calcium oxalate, cannabinol, caffeine, mustard oil, strychnine, rotenone, etc." Many of the
most familiar poisons and stimulants in human culture
come from tropical plants. These chemicals, along with
many others present in plants, are called secondary compounds because most seem to lack any direct metabolic
function.

Plants contain many different kinds of secondary compounds. Because they collectively help protect plants, they are commonly called defense compounds or allelochemics (Whittaker and Feeny 1971). They may have originated as genetically accidental metabolic byproducts or chemical wastes that, by chance, conveyed some measure of protection from either microbes or herbivores. Such mutations would have high fitness, and, thus, natural selection would favor their accumulation (Whittaker and Feeny 1971). Most researchers attribute no direct metabolic function to secondary compounds, arguing instead that they serve entirely for defense (Ehrlich and Raven 1967; Janzen 1969b). However, some evidence exists that some secondary compounds may be directly functional (Seigler and Price 1976; Futuyma 1983). Nonetheless, there is ample evidence favoring the argument that many secondary compounds function primarily for plant defense. Most plant species contain a variety of defense compounds representing several major chemical groups. Some defense compounds function principally to protect against herbivores, some to protect against bacteria and fungi.

ALKALOIDS

Alkaloids are among the most familiar and addictive drugs known. Such familiar names as cocaine (from coca), morphine (from the opium poppy), cannabidiol (from hemp), caffeine (from teas and coffee), and nicotine (from tobacco) are all alkaloids. Taken together, there are over 4000 known alkaloids that are globally distributed among 300 plant families and over 7500 species, and a single plant species may contain nearly 50 different alkaloids (Levin 1976). By no means do all alkaloid-containing species occur in the tropics. Although tropical species contain many alkaloids (see below, "Latitudinal Trends"), alkaloids are also well represented among temperate zone plants. It is estimated that approximately 20% of all vascular plant species contain alkaloids (Futuyma 1983). Alkaloids are found not only in leaves but almost anywhere in the plant, including seeds, roots, shoots, flowers, and fruits.

Most alkaloids have a bitter taste. In mammals they tend to interfere with both liver and cell membrane function. They may also cause cessation of lactation, abortion, or birth defects. Many are addictive. Of all of these characteristics, it is probably the bitter taste combined with the difficulties of digestion and liver function that discourage animals from munching alkaloid-rich vegetation.

Work conducted by James A. Nathanson (1984) indicates that caffeine can discourage insect feeding. Caffeine is a type of alkaloid called a methylxanthine, and both it and synthetic methylxanthines were shown by Nathanson to seriously interfere with enzyme systems of tobacco hornworms (*Manduca sexta*). Damage occurred at concentrations noted to be normal for field plants, thus caffeine, though a stimulant to humans, is in reality an insecticide. However, most alkaloids do not seem to function as insect antifeedants (Hubbell pers. com. 1987). Further, some alkaloids may serve not primarily as defense compounds but rather to store carbon and/or nitrogen (Futuyma 1983).

PHENOLICS AND TANNINS

Phenolic compounds are often abundant in plants (Levin 1971). One group adds the pungency to many of the most well-known spices, and another, known as the tannins, provides the basic compounds used in tanning leather. Tannins are particularly abundant in temperate oak leaves as well as in many tropical plant species such as mangroves.

Work performed on *Cecropia peltata* on Barro Colorado Island, Panama by Phyllis D. Coley (1984) indicated that tannins were heavily concentrated in young trees but declined in concentration in older plants. Tannin levels were lower in plants grown in the shade, indicating that access to high levels of photosynthesis may be necessary for tannin production. In field experiments, low tannin plants experienced twice the level of herbivory as those with high tannin levels. However, leaf production was inversely correlated with tannin levels. The more leaves on the tree, the lower the tannin per leaf, indicating that tannin production, though perhaps protective, is costly. Trees like *Cecropia*, experiencing intense competition for light, are probably best served by sacrificing tannin protection for rapid growth.

Phenolics are small proteins stored in cell vacuoles that are broken when an insect or other herbivore bites the leaf. Upon release, the phenols combine with any of the numerous proteins including those enzymes necessary

Cecropia trees often show insect damage to leaves, but cecropias grow quite quickly.

for splitting polypeptides (parts of proteins) in digestion, perhaps making it more difficult for a herbivore to digest protein. Leaf damage by insects or pathogens may stimulate production of phenolics (Ryan 1979).

By no means are phenolics or tannins effective in discouraging all insect herbivores. Leafcutting ants, major herbivores throughout the Neotropics, seem undeterred by them (Hubbell, Howard, and Wiemer 1984). The role of phenolics, and tannins in particular, as antiherbivore adaptations is questionable. There is little direct evidence, for instance, that tannins serve as a general defense mechanism (Zucker 1983). Tannins may discourage some insects initially, only to stimulate natural selection for resistance to them, as has happened, for instance, with the use of insecticides like DDT. Some insects, especially those adapted to feed on a diversity of plant foods, have enzymes that detoxify some defense compounds (Krieger et al. 1971). Some kinds of tannins may be effective principally against insects, while others may function principally against microbes and pathogens (Zucker 1983; Martin and Martin 1984).

SAPONINS

If, as a child, you were ever punished by having your mouth washed out with soap, you have some idea of how saponins taste. These soaplike compounds are relatively common in tropical plants and act to destroy the fatty component of the cell membrane. Some South American Indians utilize saponins to catch fish. Saponins, leached from leaves, act on fish gills and interfere with respiration.

CYANOGENIC GLYCOSIDES

Everyone knows that cyanide is a deadly poison. Many species of tropical plants contain compounds called cyanogenic glycosides consisting of cyanide locked together with a sugar molecule. When combined with enzymes from either the plant or a herbivore's digestive system, the sugar is released leaving hydrogen cyanide. Needless to say, these highly potent plants discourage herbivores. It is interesting to note that one of the most important food plants for humans in the tropics, manioc, from which cassava is prepared (see chapter 3, page 103), contains cyanogenic glycoside in the root, the part which is eaten. Thus the root must be washed extensively with water to eliminate the cyanide before it is eaten (see below, "Humans and Drugs").

Cyanogenic glycosides are well represented in passion-flowers (*Passiflora* spp.), fed upon by caterpillars of *Heliconius* butterflies (see below). Work conducted by Kevin C. Spencer (1984) demonstrated that each *Heliconius* species is able to detoxify one or two cyanogenic glycosides, and, thus, these butterflies are selective, each species feeding only on select passionflower species. The butterfly caterpillars have hydrolytic enzymes specific for detoxifying certain cyanogenic glycosides.

CARDIAC GLYCOSIDES

Cardiac glycosides interfere with heart function. Digitalis, from foxglove (*Digitalis purpurea*), is a cardiac glycoside used as a heart stimulant. Cardiac glycosides can be fatal to normal hearts. Among the many types of plants with cardiac glycosides are members of the milkweed family.

TERPENOIDS

Terpenoids are a complex group of fat-soluble compounds that include monoterpenoids, diterpenoids, and sesquiterpenoids. Some are used in the synthesis of compounds that may mimic insect growth hormones (preventing rather than promoting growth of the insect) or can be modified into cardiac glycosides (Futuyma 1983). Some terpenoids discourage both insects and fungi. One terpenoid in particular, caryophylene epoxide, has been shown to completely repel the fungus garden ant *Atta cephalotes* from clipping leaves of *Hymenaea courbaril* (see page 167). This terpenoid was shown to be highly toxic to the fungus that the ants culture (Hubbell et al. 1983). In a survey of forty-two plant species from a Costa Rican dry forest, three-quarters contained terpenoids, steroids, and waxes that strongly repelled leafcutting ants (Hubbell et al. 1984).

TOXIC AMINO ACIDS

Some tropical plants, especially members of the bean and pea family (legumes), contain amino acids that are useless in building protein but instead interfere with normal protein synthesis. Canavanine, for example, mimics the essential amino acid argenine. Perhaps the best known of the toxic amino acids is L-DOPA, which can be a strong hallucinogen. Both canavanine and L-DOPA have been found to be concentrated in the seeds of some tropical plants. In general, the major function of nonprotein amino acids, at least in legumes, seems to be to discourage herbivores (Harborne 1982).

CALCIUM OXALATE

Among of the most familiar tropical plants are the large-leaved philodendron vines of the arum family. The familiar skunk cabbage (*Symplocarpus foetidus*), which grows throughout swamps and streamsides in eastern North America, is among the relatively few arums to reach the northern temperate climes. Skunk cabbage, like most of the other arums, contains crystals of calcium oxalate, a caustic substance that makes the delicate tissues of the mouth burn. Most tropical arums have very large leaves, tempting targets for herbivores. Given their obvious large leaf size, it is easy to see how successfully protected arums must be. Very few of their huge leaves are chewed.

OTHER TYPES OF DEFENSE

We drive around every day on a defense strategy of tropical plants. I am, of course, referring to rubber tires. The rubber tree (*Hevea brasiliensis*), which can reach heights of 120 feet, is but one of many tropical trees that produce latex, resins, and gums, all of which act to render the trees less edible. Rubber tree sap is a milky suspension in watery liquid. It may aid in closing wounds, protect against microbial invasion, and perhaps hinder herbivores. The chicle tree, *Manilkara zapota*, which grows in rain forests throughout Central America, produces a natural latex called chicle from which chewing gum is made.

Stephen P. Hubbell and his colleagues (1984) noted that in trees whose leaves were chemically nonrepellent to leafcutting ants, sap adhesion to the insects' mouthparts and appendages discouraged ants from clipping the leaves. Leafcutter ants will take leaves that are removed from *Euphorbia leucocephala* trees but will not clip intact leaves, presumably because clipping induces copious sap flow from intact leaves (Stradling 1978).

In addition, tropical trees may be spiny, thorny, or have leaves coated with diminutive "beds of nails" called trichomes that sometimes literally impale caterpillars (Gilbert 1971). Experiments have shown that sharply toothed leaf edges can reduce caterpillar grazing. When teeth are experimentally removed, caterpillars feast (Ehrlich and Raven 1967). In Belize, I am careful to avoid grabbing the trunk of a young warree palm, *Acrocomia vinifera*. This common understory species has 2-inch-long spines lining its lower trunk. Each of these sharp, jagged stilettos is coated with lichens and various microbes and can cause infection easily if they penetrate the hand. Warree palms also have long, sharp spines on the undersides of leaf midribs. Even the actual wood of tropical trees is often quite hard, a possible adaptation to discourage the ever-present hordes of termites.

Leaf toughness, nutrition value, and fiber content also affect ability to resist herbivores. Phyllis D. Coley (1983) examined rates of herbivory and defense characteristics of forty-six canopy tree species on Barro Colorado Island. She compared young with mature leaves and gap-colonizing species with shade-tolerant species. In general, young leaves were grazed much more than mature leaves even though many were loaded with phenols (indicating that phenols do not discourage herbivory). Young leaves were, however, more rich in nutrients and lower in fiber and toughness. Coley learned that mature leaves of gap-colonizing species were grazed six times more rapidly than leaves of shade-tolerant species and that, overall, gap-colonizing trees had less tough leaves with lower concentrations of fiber and phenolics than shade-tolerant species. Gap tree leaves grew faster but had shorter lifetimes. Coley concluded that leaf toughness, fiber content, and nutritive value were more influential than defense compounds in affecting patterns of herbivory. Studies on nutritional choices of howler monkeys (see below, page 192) resulted in a similar conclusion (Milton 1979, 1981).

The typical tropical plant combines various defense compounds with mechanical defenses. Some plants contain up to ninety different defense compounds ranging from carcinogens to stimulants, laced throughout the plant's tissues. Spines may line the leaf edge, and thorns help protect the bark. Even the wood may contain crystals of silicon, a sort of "ground glass" for protection (Janzen 1975). Some plants seem to utilize insects in their defense. The bull's horn acacia, described in chapter 4, is protected by its resident *Pseudomyrmex* ants. For these acacias, ants ecologically take the place of defense compounds. Acacia species without ants contain cyanogenic glycosides, which ant acacias lack (Janzen 1966). Ants also are implicated in the defense of other species such as cecropias and passionflowers.

Latitudinal Trends

As I mentioned earlier, defense compounds are well distributed in the temperate zone and are even in the arctic, as well as in the tropics. Lemmings, the small tundra-inhabiting rodents whose populations sometimes swell enormously, are periodically affected by defense compounds in some of their food plants. It is true, however, that tropical ecosystems are habitats where defense compounds are abundantly represented. Just as species diversity increases as latitude decreases (page 123), so does the presence of defense compounds.

Donald A. Levin (1976) has made a detailed study of the geographic distribution of alkaloid-containing plants. Levin examined herb, shrub, and tree species and found that in all three groups, a significantly greater percentage of tropical species contain alkaloids. In all, 27% of temperate species tested contained alkaloids compared with 45% of the tropical species tested. The evolutionarily more primitive plants such as the magnolias, many of which are tropical species, contained more alkaloid-bearing representatives than species from the more recently evolved families.

In Kenya, 40% of the plant species tested contained alkaloids compared with only 12.3% in Turkey and 13.7% in the United States. In Puerto Rico, a tropical but much smaller geographic area than the United States, Levin found that 23.6% of the plants tested contained some alkaloids. Levin's analysis revealed a relatively steady decrease in the percentage of plants bearing alkaloids from the equator northward, a trend termed a *latitudinal cline*. One possible explanation for this cline is that nontropical species are less subject to pest pressure and thus high alkaloid content has not been selected for in many of these species. Another is that pest pressure may differ substantially between tropical and temperate areas, selecting for different defense chemicals.

Ecological Trends

Defense compounds are found in plants from virtually all habitats, but some trends are apparent. Tropical lowland forests, mangrove swamps, deserts, and mountain rain and cloud forests are all habitats where defense compounds are abundant. Alpine forest and grassland as well as disturbed areas (see below) have fewer plants containing defense compounds.

Defense compounds are abundant in lowland forest occurring on nutrient-poor white, sandy soils in the northern Amazon region (page 75). Leaves from the vegetation of white soil forests are long-lived and have such high concentrations of defense compounds that even after the leaf drops from the plant nothing can really eat it. The leaf must be leached of its defense compounds by rainfall before it can be broken down and its minerals recycled. The blackwater rivers that are characteristic of white, sandy soil regions are black because of the leached phenolics (e.g., tannins) in the water. The probable reason for the high concentration of defense compounds in white, sandy soil forests is that it is more efficient for the plants to manufacture leaves packed with defense compounds than to replace leaves ravaged by microbial pathogens, fungi, or herbivores. Given the shortage of minerals in the soil, the replacement of leaves is a more costly enterprise than the synthesis of defense compounds that aid in leaf longevity.

Areas undergoing ecological succession tend to have species that invest in defense compounds differently from those on poor soil. Most successional species are racing to maximize their rates of growth. They seem to synthesize alkaloids, phenolic glycosides, and cyanogenic glycosides, all present in low concentration, collectively representing a relatively low energy cost. In contrast, plants on nutrient poor soils invest in metabolically "expensive" defense compounds such as polyphenols and fiber (e.g., lignin) that are retained as immobile defenses in the leaves and bark. Trees grow more slowly but are well protected as they grow. Phyllis D. Coley and two colleagues (1985) have hypothesized that these contrasting patterns in plant defenses are related to resource availability. On sites where resources are poor, "expensive," long-lasting defense compounds are favored. On resource-rich sites, "cheaper," shorter-lasting defense compounds are favored because the tree is able to both devote sufficient energy to rapid growth and replace defense compounds as needed.

Many successional species are subject to herbivore damage (Brown and Ewel 1987). Cecropias are fast growing, and I have seen many with extensively damaged leaves. Perhaps cecropias "trade off" protection by defense compounds for rapid growth. I have also seen vines and herbs thoroughly munched in successional areas adjacent to mature forests where understory leaf damage was far less. Areas undergoing early succession are resource-rich. As more and more species invade and grow on the site, competition among individuals reduces the overall availability of resources per individual and selects for those species adapted to grow more slowly but persist. Daniel Janzen (1975) has estimated that insect density is five to ten times greater in successional areas than in the understory of a mature rain forest, presumably because successional species are more palatable to insects.

The battle between insect herbivores and a plant species was well documented for the understory species *Piper arieianum* at La Selva Biological Station in Costa Rica. Robert J. Marquis (1984) found that insect damage to this piper species varied greatly from plant to plant, and that some plants were genetically more resistant to herbivores than others. Marquis concluded that herbivory has had a very strong selective influence on the evolution of plant defenses. Perhaps some individual pipers "risk" investing little in defense compounds but more in rapid growth and reproduction, while others grow more slowly but are better protected. Plants with few defenses may be partially protected by occurring among well-defended plants, where herbivores are scarcer.

PLANTS VERSUS MONKEYS

Nearly everyone visiting a Neotropical rain forest for the first time places monkeys high on the list of "must see." Consider what it would be like to actually be a monkey living in a rain forest. Monkeys are mainly vegetarians and live in what would seem an idyllic habitat burgeoning with leaves and fruits. Appearances can be deceiving, however. Considering the abundance of defense compounds packed into all tissues of tropical plants, how do monkeys cope with a world green on the outside but perhaps poison on the inside?

Sometimes science begins with serendipity. A casual observation, perhaps unimportant or only vaguely interesting to most people, becomes a starting point for a study that yields new and exciting perspectives on how nature functions. For Kenneth Glander (1977), the serendipitous observation happened while he was watching mantled howler monkeys, *Alouatta villosa* (page 282), in Costa Rica. A female animal with her juvenile clutching on apparently became disoriented and fell 35 feet out of a carne asada (*Andira inermis*) tree. Monkeys do not often fall from trees. For a monkey, the tree is everything; it is food, it is shelter, it is security. Any monkey that falls from a tree is either ill or wounded. The female was dazed from her fall but able to climb back up the tree. Her baby was not injured. But, why had the monkey lost its balance and fallen out?

Glander became curious about the monkey that fell from the tree as well as about several dead howlers he found in his study area. He wondered if these animals had perhaps been poisoned by defense compounds present in the foliage they had eaten. Glander was aware of the fact that local people use the crushed leaves of the common madera negra tree, *Gliricidia sepium*, to obtain the poison rotenone, used against rats. Madera negra leaves also contain various alkaloids. Howlers eat madera negra leaves.

For Glander, the work had begun. Nature does not relinquish information easily, and Glander spent five thousand hours in the field observing howlers. He marked each of 1699 trees in the study area so that he could document the exact trees in which monkeys fed.

Glander learned that howlers were extremely selective in their feeding choices. Of a total of 149 madera negra trees in the study area, the howler troop only fed in 3, and they were always the same 3 trees! These trees were found to have leaves free of alkaloids and cardiac glycosides. Other madera negra trees had high concentrations of defense compounds. The howlers had learned which trees were safe to eat. Imagine living in a town with 149 restaurants. Most put poison in the food they serve, but 3 serve safe, healthful meals. The people of the town would learn quickly by sad experience which restaurants were safe. So it is with the howlers in Costa Rica. Glander speculated that defense compounds may have been one factor affecting evolution of intelligence in primates. Poisons provide a selection pressure to remember which trees are safe and which are not and to communicate the information to others in the troop. The very social structure of the howlers and other monkey species may be evolutionarily related to stresses imposed by living in a world of toxic trees.

Glander found that mantled howlers tended to favor young leaves that are relatively high in nutritional value but have not yet become loaded with defense compounds. When given no choice but to eat mature leaves, they ate only a little and then moved to a different tree. This behavior would help avoid too high a dose of one type of defense compound. Sometimes they ate only the leaf stalk or petiole, ignoring the blade. The petiole has the least alkaloid content. The total picture is not that simple, however. Katherine Milton (1979, 1982), who studied mantled howlers on Barro Colorado Island in Panama, found that protein and fiber content, not secondary compounds, seemed to be more important factors affecting leaf choice. Fiber makes leaves more difficult to digest and is least concentrated in the preferred younger leaves. Protein is proportionally higher in young than older leaves. The more protein to fiber, the more desirable the leaf was to howlers.

Milton (1981) learned that howlers have very long intestinal systems, especially the hindgut. It takes food up to 20 hours to pass through a howler's digestive system. Spider monkeys, *Ateles geoffroyi*, eat mostly fruits, much easier to digest because they contain more protein and fewer defense compounds than leaves. It takes food only about 4.4 hours to move through the shorter gut of a spider monkey. Howlers, with their long hindguts, are able to efficiently digest leaves, coping to a reasonable degree with both high fiber and defense compounds. Howlers and spider monkeys do not compete intensively with each other for food because each focuses on a different primary food source.

Plants versus Insects

Given that tropical plants contain large quantities of defense compounds, how do insects manage to find anything to eat? One answer, at least for some, appears to be evolutionary specialization, adapting to eat relatively few kinds of plants. Unlike most monkeys and other large herbivores with more catholic diets, many insects, both tropical and temperate zone, have tendencies to feed exclusively on one type of plant rather than upon a wide array of species. For instance, heliconid butterfly caterpillars, which I discuss below, feed exclusively on vines in the genus *Passiflora*, the passionflowers. The colorful caterpillar of *Pseudosphinx tetrio* (page 118) feeds only on the leaves of *Plumeria rubra*, the frangipani tree. In Trinidad, I observed at least a dozen of these large, boldly patterned caterpillars totally defoliate a frangipani, while all surrounding trees were untouched.

Toxic defense compounds are probably one major selection pressure in the evolution of specialization among insect herbivores. Those insects that evolve enzyme systems that detoxify defense compounds or somehow sequester them (see below) are able to focus their appetites on specific plant species, eating them with abandon. Futuyma (1983) points out, however, that other factors can also select for diet specialization. For instance, plant compounds may be repellent but not actually be toxic. Insects may overcome the repellency and adapt to "recognize" a host plant by its "repellent" compounds. Interspecific competition may also select for host specialization among insects (page 126), as might avoidance of certain parasitoids (which are attracted to certain plant-specific com-

pounds). Insects might also be practicing optimal foraging behavior, searching for the food that provides the
highest reward for the smallest effort.

Many tropical insects are not specialized to feed on but
a single type of host but are generalist feeders. For example, leafcutter ants are among the least specialized herbivores in the tropics, typically foraging on a wide diversity of plant species (Rockwood 1976; Hubbell et al.
1984).

Bruchid Beetles

Beetles in the family Bruchidae provide examples of
both feeding specialization and adaptation to host defenses. Bruchids are seed predators, especially on legumes. Females lay eggs on seed pods or directly in seeds.
Larvae enter pods upon hatching (or are hatched inside
pods) and feed on seeds before pupating. Daniel Janzen
(1975) observed that one bruchid species, *Merobruchus co-
lumbinus*, killed 43% of the seed crop from its host tree,
Pithecelobium saman in Guanacaste, Costa Rica. Janzen also
noted that in Costa Rica, 102 of 111 species of bruchid
beetles from a deciduous forest feed on only one host
plant. In a survey of the beetle families Bruchidae, Cur-
culionidae, and Cerambycidae, Janzen (1980c) found
about 75% of the species fed on a single plant species.
Only 12% fed on three or more host plants.

Bruchid species are widespread in the tropics, and a
single species of bruchid may feed on several different
host plants throughout its range. However, Janzen found
that within any local area a bruchid species will feed only
on one host species. In other words, local races exist
within some bruchid species, each of which is specialized.
Bruchids have a diverse array of adaptations for dealing
with the defenses of their host plants (Janzen 1969b).

Ted Center and Clarence Johnson (1974) have documented the complex game of coevolution between bruchids and their host plants. Some host plants have
evolved toxic seeds containing hallucinogenic alkaloids,

saponins, pentose sugars, and free amino acids. Bruchids either avoid these toxins or are physiologically resistant to them (see below). Some seed pods produce sticky gum following penetration of bruchid larvae. Bruchids evolved a period of quiescence in embryonic development, until seeds mature and it is too late for pods to produce the gum. Some pods fragment or "explode," scattering seeds and thus avoiding larvae that enter after hatching on the pod wall. Bruchids evolved the habit of ovipositing directly on the seeds, after they are scattered. Some seed pods flake, thus helping remove eggs from their surface. Bruchids oviposit beneath the flaking substance, thus avoiding it, or else they have an accelerated embryonic development, hatching and entering the pod before it flakes. Some seeds remain very small (thus not supplying much bruchid food), then growing quickly just prior to dispersal. Bruchids nonetheless enter and eat immature seeds, or else delay their maturation until seeds are mature and thus larger. Finally, some legumes make very tiny seeds, too small for bruchids to mature within them. Bruchids eat them anyway, devouring several seeds during development, not just one. Both Janzen (1969b) and Center and Johnson (1974) conclude that bruchids adapt sufficiently well that most host plants are rarely free from bruchid seed predation.

The physiological ability to deal with plant toxins may be fortuitous, at least in the case of bruchids and other beetles that digest seeds containing the toxic amino acid L-canavanine. John A. Bleiler, Gerald A. Rosenthal, and Daniel Janzen (1988) found that certain species of beetles that specialize on eating seeds high in L-canavanine contain an arginase enzyme that can efficiently degrade the toxic amino acid. However, many beetles that do not feed on plant seeds high in L-canavanine also have arginase. Conceivably, these species could also detoxify L-canavanine. Why don't they? Degradation of L-canavanine may still result in its incorporation into aberrant proteins, called canavanyl proteins. Other metabolic systems are also necessary to efficiently deal with L-canavanine, which thus far, these other insects have probably not evolved.

An analysis of larvae of two beetle species that dine on L-canavanine-containing seeds found that both could cleave L-canavanine to L-canaline and urea. Rather than excrete the urea, which would represent a significant loss of nitrogen to the developing insect, the insects utilize another enzyme, urease, to synthesize ammonia from urea. The ammonia is incorporated into synthesizing various amino acids used in growth, and the L-canaline is not incorporated into aberrant amino acids (Bleiler et al. 1988).

Heliconid Butterflies and Passionflowers

Butterflies in temperate and tropical areas are known for their affinities for feeding on specific plant families. The pipevine swallowtail caterpillar (*Battus philenor*) of temperate woodlands feeds on nothing but plants of the birthwort family (Aristolochiaceae). The painted lady caterpillar (*Vanessa virginiensis*) is hooked on composites (daisies, etc., Compositae), and the monarch caterpillar (*Danaus plexippus*) specializes on milkweeds (Asclepiadaceae). Caterpillars are more selective than adults because adult butterflies feed on nectar, aiding in pollen dispersion. Their involvement with plants is fundamentally mutualistic, and defense compounds would be of less selective importance. Caterpillars, however, are voracious herbivores, and, being folivores, they encounter high concentrations of defense compounds as they chew on leaves. Since different families of plants produce different combinations of defense compounds, natural selection has acted on caterpillars in such a way that some different caterpillar species have evolved tolerance for defense compounds associated with a particular plant family.

Heliconid butterflies (Heliconiinae) are a diverse and colorful group, almost all of which are Neotropical (DeVries 1987). They are usually considered part of the brush-footed butterflies (Nymphalidae), although some

Heliconid butterfly.

taxonomists place heliconids within their own distinct family. Nymphalids number nearly three thousand species globally but Heliconids are represented by only about fifty species with many local races throughout tropical America. Commonly called longwing butterflies, only two species, *Heliconius charitonius* and *H. erato*, regularly reach the United States, and both of these are found only in the extreme southern part of the country.

Heliconid caterpillars feed almost exclusively on species of *Passiflora* or passionflower (Passifloraceae), a common vine numbering approximately five hundred species. Like heliconids, passionflowers are largely Neotropical, and another name for heliconids is "passionflower butterflies" (DeVries 1987). Not many kinds of herbivores eat passionflower vines, probably because these vines contain effective defense compounds. A study by Kevin C. Spencer (1984) revealed that most passionflowers contain various cyanogenic glycosides and cyanohydrins. Further, Spencer documented a strong correlation between preferences by heliconid caterpillars and specific defense compound present. The high diversity of cyanogens among passionflower species may be an evolutionary response to herbivory by heliconids. Heliconids, however, seem able to adapt to the ever-changing cyanogen regime by evolutionary changes in their hydrolytic enzymes and by sequestration of cyanogens. Heliconid caterpillars have largely solved whatever toxic challenges may be presented by passionflower and are capable of defoliating passionflower with little difficulty. To passionflower, these caterpillars are the enemy.

Heliconid butterflies lay small numbers of eggs in globular yellow clusters directly on passionflower leaves, favoring young shoots. When the eggs hatch, the little caterpillars are sitting on their food source. The trick for the plant, therefore, is to prevent the adult female butterfly from locating, selecting, and laying eggs on its leaves.

Detailed studies (Gilbert 1975, 1982; Benson et al. 1976) of passionflower versus heliconiines have shown that the plant's defenses seem to be both an attempt to repel and to "fool" the caterpillars. Passionflower produces extrafloral nectaries (page 163) that attract various species of ants and wasps. These aggressive insects help repel heliconid caterpillars. The chemistry of the extrafloral nectaries was studied by Jonathan M. Horn (1984) and two colleagues, and they learned that certain amino acids such as proline, tryptophan, and phenylalanine were very common constituents of nectaries. Proline is important for insect flight, and the other two amino acids are important in insect nutrition. The nectaries attract flying hymenopterans (bees and wasps) as well as ants. Some passionflowers are protected exclusively by ants, some exclusively by wasps, some by both. John T. Smiley (1985) conducted an experiment at Corcovado National Park in Costa Rica to test the efficacy of ants in discouraging heliconid caterpillars. He found that caterpillar survival is much lower on *Passiflora* with attending ants. Caterpillar mortality rate was 70% on ant-attended plants compared with 45% on nonant plants. Smiley concluded that ants do protect passionflowers from heliconids.

Lawrence E. Gilbert has noted that some *Passiflora* extrafloral nectaries appear, to the human eye, to mimic heliconid egg clusters. Passionflower vines typically have young leaves spotted with a few conspicuous yellow globs, the egg mimics. Female heliconids will not lay eggs on a leaf already containing egg masses. The "mimic" egg masses presumably prompt the female to "keep searching" and find another victim. Gilbert believes the mimic eggs to be a recent evolutionary development in the plant-insect battle. Currently only 2% of passionflower species have mimic eggs.

Leaf shape varies within a species of passionflower. Gilbert notes that passionflower leaves often resemble other common plant species growing nearby. He hypothesizes that heliconids may be tricked by the similarity of appearance (leaf mimicry) and thus overlook a passionflower. This speculation depends, of course, on the butterfly using visual cues to locate passionflower vines. If the insect depends principally on scent, Gilbert's idea could be weakened. More research is required to answer the question.

For the time being, at least one passionflower species, *Passiflora adenopoda*, may have won the coevolutionary battle between insect and plant. Its leaves are covered by minute hooked spines called trichomes. Resembling a Hindu's bed of nails, trichomes impale the soft-skinned caterpillars. Once a caterpillar is stuck, it starves (Gilbert 1971). Does the future hold in store thick-skinned caterpillars? Time will tell. At least one butterfly species, *Mechanitis isthmia* (an ithomiid, not a heliconid), has adapted to thwart trichome defense. Caterpillars feed on plants of the tomato family and succeed in countering trichomes by spinning a fine web covering the trichomes, enabling the caterpillars to move easily over the leaf surface to feed on the leaf edges (Rathcke and Poole 1975).

Why Are Heliconius Butterflies Pretty?

Heliconid butterflies are neither rare nor inconspicuous. In fact, they are among the most obvious and beautiful butterflies of the tropics. They fly slowly, almost delicately and are very easy to see along forest edges and well-lit disturbed areas. When atop a plant, the butterfly appears brilliantly colored, many almost iridescent. Why are they so conspicuous? Consider the potential risk to the insect. Most tropical bird species feed heavily on insects. Remember, there are over three hundred species of flycatchers alone! It would seem suicidal for a butterfly group to be colored like neon signs, seeming to say "eat me."

Remember, however, that in the tropics, obviousness may serve as a warning (page 116). If heliconid butterflies are distasteful to would-be predators, it may be to the butterflies' potential advantage to be obvious. Once a bird has eaten an unpalatable insect, it may remember its unpleasant experience and steer clear of other insects in the group.

Heliconid butterflies are only one group among many that are brilliantly colored, including one well-studied example from the temperate zone. The conspicuous orange monarch (*Danaus plexippus*), familiar to anyone even vaguely interested in lepidopterans, is another example. Lincoln Brower (1964, 1969) tested the palatability of monarchs to blue jays (*Cyanocitta cristata*) and learned that monarchs are poisonous to the jays. Minutes after a jay eats a monarch, it vomits up the remains. From that point on, the bird will not touch a monarch. The bright color pattern of the butterfly is remembered by the bird.

Monarch caterpillars feed on milkweeds, many of which contain high concentrations of cardiac glycosides. Caterpillars sequester glycosides in their tissues, and after metamorphosis the adult butterfly contains glycosides, rendering it unpalatable. Monarchs have turned a neat evolutionary trick. They have adapted to the milkweed's defense compounds, cardiac glycosides, and turned it to their own protection. Not only do they obtain virtual exclusive use of milkweed (few other insects can eat it), but they also are protected by using milkweed's defense compounds. It is possible that heliconid species do the same thing (Brower et al. 1963). Butterflies adorned by warning coloration are likely to store plant defense compounds and thus to be able to sicken their predators (Brower and Brower 1964).

In Costa Rica, Woodruff Benson (1972) altered the coloration pattern of *Heliconius erato*, a common, brightly colored, unpalatable butterfly. He marked the butterflies to alter their wing patterns such that predatory birds would not have seen such patterning before. He also established a control group, identical to specimens normally occurring. He released equal numbers of normal "control" butterflies and altered, uniquely patterned individuals. Significantly fewer of the altered individuals were recaptured, an indication that fewer of these survived. Some altered butterflies that were recaptured showed damage from bird attacks. Benson was even able to identify one bird species, the rufous-tailed jacamar (*Galbula ruficauda*), by the shape of wounds it left in a butterfly's wing. Benson concluded that wing pattern does indeed serve as protection, once predators learn the pattern and associate it with unpalatability.

There is an obvious cost associated with warning coloration. At each generation some individuals are "sacrificed" in order to "educate" the predators. Why should a butterfly give up its life for the good of its species? The answer is that it probably doesn't but, rather, gives up it's life only for the good of it's own genes. Heliconid butterflies have a limited home range and a single female can live for up to six months, a virtual Methuselan age for a butterfly. Because the basically sedentary females lay many eggs over the course of their long lives, the local population in any given area is likely to consist of close relatives: brothers, sisters, cousins, second cousins, etc. If a single individual is lost to a predator, but the predator subsequently avoids other members of the group, the other group members each benefit. And, if most of the members of the group have genes in common with the sacrificed individual, then, in an evolutionary sense, the individual has indirectly promoted its own genetic fitness. It has acted to protect *copies* of its own genes. This process, called kin selection, may help explain how warning coloration evolved in heliconids. Thus far, it has not been rigorously tested and so must remain speculative. Interestingly, heliconid butterflies are among the only butterflies to have communal roosts, aggregations that may be "extended families".

MIMICRY SYSTEMS

Although heliconids and many other tropical butter-flies are striking in appearance, many are often difficult to identify because many species look remarkably alike: they mimic each other. When you try to identify a tropical butterfly, look very closely because it may not be what it appears to be. It may be another species entirely, a mimic of what you think it is.

Batesian Mimicry

Henry Walter Bates (1862), one of the first naturalists to explore Amazonia, discovered that some unrelated species of butterflies look alike. He suggested that a pal-atable species may gain protection from predators if it closely resembles a noxious unpalatable species, a phe-nomenon now termed Batesian mimicry. Bates was cor-rect. The unpalatable species, termed the model, is essen-tially parasitized by the palatable species, termed the mimic. Because it closely resembles an unpalatable spe-cies, the mimic enjoys the umbrella of protection pro-vided by the presence of the model. For the model, the presence of the palatable mimics makes the education of predators more difficult. Suppose a predator encounters one or even two palatable mimics *as its first experience*. It may be subsequently more difficult for the predator to "believe" that the noxious model is always noxious. Bate-sian mimicry is most effective when the mimic is not very abundant. If, for instance, it were as abundant as its model, the entire system would be relatively unprotective, because predators would encounter palatable mimics as readily as unpalatable models.

A classic example of Batesian mimicry is the North American viceroy butterfly, *Basilarchia archippus*. The viceroy, a member of the family Nymphalidae, bears a strong resemblance to the monarch, a Danaid. Both are orange with black wing veins. As I discussed above, monarchs, by virtue of a diet tainted by cardiac glycosides from milkweeds, are unpalatable. Viceroys feed mostly on plants of the willow family and are quite palatable. Any lepidopterist will tell you that it's much harder to find a viceroy than a monarch. The mimicry works but only if the mimic's population is small in any given area. (One interesting twist on mimicry is that some monarchs "mimic" themselves! There are milkweeds that contain little or no cardiac glycosides and are fed upon by monarch caterpillars. Butterflies from these caterpillars are palatable, but still gain protection by occurring with monarchs that are unpalatable, a phenonmenon Lincoln Brower [1969] terms *automimicry*.)

Viceroys, especially in the southern part of their range, also mimic another milkweed-feeding butterfly, the queen (*Danaus gilippus*). Some tropical butterfly species actually mimic several different models. By utilizing more than one model, the mimic's total population may be larger.

Butterflies exhibit many examples of Batesian mimicry in the Neotropics (Gilbert 1983). However, they are not the only Batesian mimics. Paul A. Opler (1981) has unraveled a Batesian mimicry system in which a single innocuous insect species, *Climaciella brunnea*, a predatory mantispid in the order Neuroptera (to which belong the lacewings, alderflies, dobsonflies, and others), has evolved an uncanny resemblance to five different wasp species. The stingless neuropterans, like viceroys and other butterflies, mimic an array of models, and all gain protection by appearing to be what they are not.

Müllerian Mimicry

Although Batesian mimicry is well represented in the tropics, recall that most plant species there have some combination of defense compounds. Therefore, any caterpillar species will likely have to cope with defense compounds in adapting to a food source. Unpalatability of caterpillars should be relatively common in the tropics, because so many of the food plants encountered by the larval insects have defense compounds that, if stored by the insect, could render the creature unpalatable. In 1879, Fritz Müller suggested that two or more unpalatable species could benefit by close resemblance. If two unpalatable species look alike, the would-be predator only needs to be educated once, not twice. The closer the resemblance, the greater the advantage to both species. This concept of convergent patterns among unpalatable species is called Müllerian mimicry. Unlike Batesian mimicry, Müllerian mimicry is basically mutualistic, because individuals of both species benefit from the mimicry.

Both *Heliconius erato* and *H. melpomene* are unpalatable, both are very brilliantly colored, and they look alike. What is remarkable is that there are eleven distinct races of *H. melpomene* in the American tropics ranging from Mexico to southern Brazil. These races do not look the same. John R. G. Turner (1971, 1975, 1981) learned that for every local race of *H. melpomene*, there is a virtually identical local race of *H. erato*. Both species have converged in *racial variation* throughout their ranges! Only one race of *H. erato*, which is restricted to a small range in northern South America, lacks a *H. melpomene* counterpart. These two species provide a clear example of Müllerian mimicry.

In the southern United States and Caribbean, another example of Müllerian mimicry is the resemblance between the monarch and queen butterflies, both of which feed on milkweeds and, thus, are both unpalatable (Brower 1969).

Mimicry Complexes

Both Batesian and Müllerian mimicry have led to extensive similarities among large groups of butterflies in the tropics. Dozens of distinct species converge in appearance making butterfly identification a taxomomist's nightmare. Christine Papageorgis (1975), working in Peru, has identified five distinct color complexes of butterflies, each of which contains many mimicking species. She also found that each mimicry complex tended to occupy a distinctly different height in the rain forest.

From the forest floor to 2 meters is the *transparent complex*, butterflies with transparent wings with black wing veins. From 2 to 7 meters is the *tiger complex*, consisting of yellow, brown, black, and orange striped butterflies, most of which are Müllerian mimics. From 7 to 13 meters, the *red complex* dominates. These butterflies include both *Heliconius erato* and *H. melpomene*. From 15 to 30 meters is the *blue complex*, all of which are heliconids with irridescent blue on the hindwing and yellow bands on the forewing. Finally, from 30 meters to the canopy and above is the *orange complex*, another group of heliconids with bright orange wings with black wing veins.

Papageorgis wondered why these five complexes were vertically stratified in the rain forest. After some experimental work, she concluded that each complex may be relatively cryptically colored in flight at the height in the rain forest where it normally flies. She related the coloration pattern of the complex to the pattern of light penetration in the rain forest and suggested that each color complex is most difficult for predators to see *in flight* at the height normally occupied by the complex. Papageorgis argued that brightly marked butterflies do not necessarily attract predator attention but, "on the contrary, can be cryptic in flight, thus giving a butterfly pursued by an uninformed or forgetful predator a second chance of escaping into the foliage." If Papageorgis is correct, tropical butterflies may employ both cryptic and warning coloration within the same species, depending upon where, exactly, they are flying, or if they are perched.

HUMANS AND DRUGS—FROM CASSAVA TO BALCHE

Native peoples in the tropics have had long experience in dealing with defense compounds. Manioc (page 103), whose thick root provides a basic source of carbohydrate, is protected by cyanogenic glycosides. An energy-rich root growing in tropical soil would not stand much of a chance against subterranean insects were it not for its toxicity. Interestingly, some forms of manioc have relatively little cyanide (Hansen 1983b). Those that grow in the most fertile soils, have a white root with a sweet taste. The poorest, least fertile soils grow manioc with the highest concentrations of cyanide. These roots are yellowish and taste bitter because of the cyanide. Just as with most plants on infertile white sand soils (page 75), manioc abounds with defense compounds. Because of its cyanide protection, manioc is one of the few crops that people grow successfully on poor soils.

When manioc is heavily laced with cyanide, it must be repeatedly washed and then cooked to remove the toxin. In Belize, women employ a 5-foot cylindrical cone, woven from palm fronds, that constricts when pulled and is used to "wring out" dough as cassava is prepared from manioc root. This cone is named a *whola*, the local name for the boa constrictor.

Other than fruits, the tissues of few species of tropical vegetation are edible by people. Most of the uses that native peoples have made of secondary compounds have been related to medicinal effects (see below) or their effects as intoxicants and hallucinogens. Aztecs and Mayans both used mushrooms and various "psychedelic fungi" in their religious rituals. One mushroom in particular was said to provide its users with "visions of Hell" (Furst and Coe 1977). *Riviea corymbosa*, a large, white morning glory, was used by the Mayans as a hallucinogen at the time of the Spanish invasion of Central America. The hallucinogens are alkaloids related to LSD. Another widely used alkaloid hallucinogen is peyote, made from fermenting cactus or agave.

One heavily used intoxicant was called balche. Mayans used balche in many of their religious rituals. It was made from fermented honey plus a bark extract from a tree, *Lonchocarpus longistylus*. When ingested by mouth, balche, as well as other potent drugs, often caused the user to become violently ill with nausea. Peter Furst and Michael Coe (1977) have suggested that Amerindians may have routinely taken these drugs by rectal injection, a practice they refer to as ritual enemas. Illustrations on Mayan vases depict both the self-administration and assisted administration of enemas. Even the Mayan gods were so depicted. It is unlikely that such an event would have been singled out by an artist if the enema was being taken to relieve constipation. Furst and Coe believe that the widespread use of enemas was part of important Mayan ritual and that, by this practice, the people avoided the nausea that would occur if the balche had to pass through the stomach. They believe the practice to have been wide-spread and to have occurred among Aztec and Incan peoples as well as Mayan.

On June 19, 1979, I met Piwualli, a bush doctor whose circuit includes the villages along a section of the Napo River in Ecuador. Said to be seventy-three but looking at least twenty years younger, Piwualli ushered my group to a dilapidated wooden table outside his house. Neatly placed on the table were clumps of dried herbs, the bush doctor's pharmacy. Piwualli spoke no English and only a little Spanish, but Enrique, our guide, knew Piwualli's native Indian language and acted as translator. With his six children, two yapping dogs, and one screeching parrot providing background sounds, Piwualli would pick up one of the herb clumps and, with a look of wisdom on his weathered face, tell us for what it was used. Enrique translated, "This one is for the headache. And this one is for the constipation. This one is for fever. This is for dysentery. And this one is for the Parkinson's disease." Parkinson's disease? I was incredulous at the thought that a bush doctor in Ecuador might be able to treat Parkinson's disease with a clump of dried plants. Frankly, I still am. Enrique told my group that Piwualli had described the symptoms for which the plant was used during the visit of another group that happened to include a physician. The M.D. said it sounded to him like a description of Parkinson's disease and supposedly took some of the herb back to the United States for testing. I have yet to read of a tropical herb cure for Parkinson's, so perhaps something was lost in the translation. Nonetheless, the possibility of such a lucky find should not be excluded.

There are many unknown drugs present in tropical vegetation. Considering how little is known about defense compounds and their effects, the potential for medical uses is vast. The National Research Council reported that approximately 260 kinds of plants are ultilized by the various peoples of the Amazon to control fertility (Myers 1980). To what additonal uses might drugs from tropical species be put?

Norman Myers (1984) cites surveys estimating that approximately 1400 plant species from tropical forests may contain drugs potentially useful in treatment of various cancers. Myers further points out that only 1 in 10 tropical forest plants has been screened in a cursory way for anticancer properties and only 1 in 100 has been seriously examined. Myers concludes:

> In Latin America alone, with its estimated 90,000 species, we have tested no more than 10,000 for their anticancer properties; and our experience to date suggests that of the remaining 80,000 at least 8,000, should reveal some form of activity against laboratory cancers, and of these, 3 could eventually rank as "superstar" drugs.

The sad fact is that in spite of all the promise shown for finding potent drugs of medical use from the tropics, tropical forests in many areas are being destroyed at such a rapid rate that we may never know what we missed (see chapter 10).

CHAPTER 6

Neotropical Birds

WHEN Henry Walter Bates (1892) was exploring Amazonia he was moved to comment on the difficulty of seeing tropical birds in dense rain forest:

> The first thing that would strike a new-comer in the forests of the Upper Amazons would be the general scarcity of birds: indeed, it often happened that I did not meet with a single bird during a whole day's ramble in the richest and most varied parts of the woods. Yet the country is tenanted by many hundred species, many of which are, in reality, abundant, and some of them conspicuous from their brilliant plumages.

The apparent scarcity of birds in tropical forests seems surprising because more species of birds occur there than in any other kind of ecosystem. Entire families, including cotingas, manakins, toucans, woodcreepers, antbirds, and ovenbirds, are essentially confined to the Neotropics, as are such unique species as screamers, trumpeters, sunbittern, hoatzin, and boat-billed heron. Bates put his finger on the irony of bird-watching in the tropics. Even birds with glamorous plumages can be remarkably silent, still, and difficult to spot in the dense, shaded foliage. Patience, persistence, and keen eyes are required of the tropical birder. Birds often seem to appear suddenly, because a dozen or more species may be moving together in a mixed species foraging flock (page 50), and, thus, the bird-watcher faces a "feast or famine" situation. One minute birds seem absent. Then suddenly they are everywhere. Bates described such an encounter:

> There are scores, probably hundreds of birds, all moving about with the greatest activity—woodpeckers and Dendrocolaptidae (from species no larger than a spar-

row to others the size of a crow) running up the tree trunks; tanagers, ant-thrushes, humming-birds, fly-catchers, and barbets flitting about the leaves and lower branches. The bustling crowd loses no time, and although moving in concert, each bird is occupied, on its own account, in searching bark or leaf or twig; the barbets visiting every clayey nest of termites on the trees which lie in the line of march. In a few minutes the host is gone, and the forest path remains deserted and silent as before.

In this chapter, I survey Neotropical birds, their adaptations, and basic ecology. Perhaps the most notable characteristic of the Neotropical avifauna is its extreme species (and subspecies) diversity (Haffer 1985). For a complete list of species, see Meyer de Schauensee (1966) or Howard and Moore (1980). Austin (1985) has provided a concise guide to bird families and Austin (1961) provides a somewhat dated but still very useful survey of the world's birds, including, of course, the Neotropics. Perrins and Middleton (1985) provide concise general natural history information on the world's birds. For detailed life histories of selected Neotropical species see Skutch (1954, 1960, 1967, 1969, 1972, 1981, 1983). David Snow (1976, 1982) has written two books focusing on his studies of frugivorous birds and has provided numerous species accounts in the professional literature. A comprehensive volume edited by Buckley et al. (1985) is invaluable for the serious student of Neotropical ornithology. For birders interested in identification, I include a list of regional field guides in the references.

The dark complex foliage of interior rain forest hosts the majority of tropical bird species. From forest floor to canopy, hundreds of different species probe bark, twigs, and epiphytes for insects and spiders. Others swoop at aerial insects, follow army ants as they scare up prey, search for the sweet rewards of fruit and flowers, or capture and devour other birds, mammals, and reptiles. One bird, the harpy eagle (*Harpia harpyja*), stalks monkeys and sloths. Patience and luck are needed to see birds well, especially when they may be 100 feet or so above ground moving through dense vegetation.

Tropical "Chickens and Turkeys"

TINAMOUS

Even ground dwellers can be elusive. Forty-four species of tinamous comprise the order Tinamiformes, a peculiar group of birds endemic to the Neotropics. A tinamou is a chickenlike chunky bird with a short, slender neck, a small, dovelike head, and thin, pointed beak. Plumage ranges among species from buffy to deep brown, russet, or gray, often with heavy black barring. Some tinamous inhabit savannas, pampas, and mountainsides, but most live secretive lives on the rain forest floor, searching for seeds and an occasional arthropod. Forest tinamous are

Tinamou.

more often heard than seen. One of the most moving sounds of the rain forest is that of the great tinamou (*Tinamus major*), a clear, ascending, flutelike whistle given at dusk, a haunting sound that heralds the end of the tropical day. One bird begins and soon others are calling. Evening twilight is the hour of the great tinamou chorus— they rarely sing during full daylight or dawn (Skutch 1983). Basically solitary, the function of the tinamou chorus may be to signal each other as to their various whereabouts.

The best way to see a tinamou is to quietly walk along a forest trail at dawn. You may suddenly come upon one on the trail, and it will probably scurry into the undergrowth when it detects your presence. Tinamous are generally reluctant to fly but may abruptly flush in a burst of wings, landing but a short distance away.

Though superficially resembling chickens, tinamous are most closely related to ostriches, rheas, and other large flightless birds. They are considered as both an ancient and anatomically primitive group. Their rounded eggs are unusual for their highly glossed shells and range of colors, from turquoise blue and green, to purple, deep red, slate gray, or brown. Only the male incubates the eggs (Skutch 1983).

CHACHALACAS, GUANS, AND CURASSOWS

Chachalacas, guans, and curassows (family Cracidae) are, like chickens and turkeys, members of the order Galliformes. They are found in dense jungle, mature forest, montane, and cloud forest. Though often observed on the forest floor, small flocks are often seen perched in trees. All species nest in trees. Delacour and Amadon (1973) provide a detailed overview of the cracids plus individual life history accounts.

The nine chachalaca species are each slender, brownish olive in color, with long tails. Each species is about 20 inches from beak to tail tip. A chachalaca has a chicken-like head with a bare red throat, usually visible only at close range. Most species form flocks of up to twenty or more birds. The plain chachalaca (*Ortalis vetula*) is among

Female great curassow.

the noisiest of tropical birds. Dawn in the jungle is typically greeted by a host of chachalaca males, each enthusiastically calling its harsh and monotonous "cha-cha-lac! cha-cha-lac! cha-cha-lac!" The birds tend to remain in thick cover, even when vocalizing, but an individual may call from a bare limb, affording easy views.

Twenty-two species of guan and thirteen species of curassow occur in Neotropical lowland and montane forests. Larger than chachalacas, most are the size of a small, slender turkey with glossy, black plumage set off by varying amounts of white or rufous. Some, like the horned guan (*Oreophasis derbianus*) and the helmeted curassow (*Pauxi pauxi*), have bright red "horns" or wattles on the head and/or beak. Others, like the blue-throated piping guan (*Pipile pipile*) have much white about the head and wings and a brilliant patch of bare blue skin on the throat.

Guans and curassows, though quite large, can be difficult to observe well. Small flocks move constantly in the

Razor-billed curassow.

canopy, defying you to get a satisfactory binocular view of them. Like chachalacas, guans and curassows are often quite vocal, especially in the early morning hours.

Both New World turkey species occur in the tropics. The widely domesticated common turkey (*Meleagris gallopavo*), which graces the Thanksgiving table with its cooked presence, once ranged south to Guatemala. Now only domesticated individuals are found throughout the tropical portion of its range. The spectacular ocellated turkey (*Agriocharis ocellata*) ranges from the Yucatán south through Guatemala. Smaller than the common turkey, the ocellated has a bright blue bare head with red tubercles. Its plumage is more colorful than its relative, particularly its tail feathers, which have bright blue and gold eyelike markings that give the bird its name. Ocellated turkeys are most easy to see at Tikal National Park in eastern Guatemala.

The Gaudy Ones

Several groups of Neotropical birds are known for their bright colors. Among them are the trogons, mot-

...udy Neotropical birds. Resplendent quetzal (far left), ...fous motmot (center), white-tailed trogon (upper right), ...ck-chinned jacamar (middle right), keel-billed toucan ...wer right).

mots, toucans, cotingas, manakins, parrots, and tanagers. The ecology of cotingas, manakins, parrots, and tanagers is intimately associated with their diets of fruit. Trogons, motmots, and toucans eat more varied diets and are among the largest birds of the jungle.

TROGONS

There are thirty-seven species of trogons (Trogoniformes) found in the world's tropics and subtropics, and twenty-three are Neotropical. A trogon has a recognizably distinct shape: a chunky, squarish bird with a long, rectangular tail and short, chickenlike bill. Brilliantly colorful, males have iridescent green and blue heads and backs, and bright red or yellow breasts. Females resemble males but are duller in color. The pattern of black, gray, and white on the tail and the color of the eye-ring (a patch of colorful skin circling the eye) are important field marks to identify various species. They range in size from about 9 to 15 inches.

Trogons tend to sit upright with tail hanging vertically down. They remain still, and it is commonplace to overlook them. The easiest way to spot one is to look for its swooping flight, flashing the bird's bright plumage, and note where it lands. Most trogons vocalize throughout the day, often a repetitive "cow, cow, cow," or "caow, caow, caow." Sometimes the note sounds harsh, but in some species it's softly whistled and melodious. A good way to see a trogon at close range is to try to imitate its call. If the imitation is good, trogons will "come in" to investigate. Some species are common along rain forest edges or successional areas. Look for their characteristic upright shape perched in cecropia trees.

Trogons are cavity nesters. Some species excavate nest holes in decaying trees, others dig into termite mounds. The violaceous trogon (*Trogon violaceous*), common from Mexico throughout Amazonia, utilizes large wasp nests. Alexander Skutch (1981) observed how a pair of violaceous trogons "took over" a wasp nest. The pair excavated their nest over several days in the cool early morning hours before the wasps became active. Skutch observed the trogons attack the wasps.

> Perching farther from the vespiary than they had done while watching each other work in the cool early morning, they made long spectacular darts to catch the insects, sometimes seizing them in the air, sometimes plucking them from the surface of their home. A sharp tap rang out each time a trogon's bill struck the vespiary in picking off a wasp.

The two trogons never eliminated all of the wasps but did successfully nest, snapping up fresh wasps daily. Oddly, few wasps attempted to sting the trogons or drive them away.

Trogons feed on fruits from palms, cecropias, and other species, which they take by hovering briefly at the tree, plucking the fruits. They also catch large insects and occasional lizards, swiftly swooping down on them or snatching them in flight. Trogon bills are finely serrated, permitting a tight grip on food items.

The most spectacular member of the trogon family is the resplendent quetzal (*Pharomachrus mocinno*), which appears on Guatemalan currency and is said to be the inspiration for the legendary phoenix. Roger Tory Peterson and Edward L. Chalif (1973) describe the quetzal as "The most spectacular bird in the New World." The male is "intense emerald and golden green with red belly and under tail coverts" (Peterson and Chalif 1973). The male's head has a short, thick crest of green feathers and a stubby, bright yellow bill. Most striking are the male's upper tail coverts, graceful plumes that stream down well below the actual tail, making the bird's total length 24 inches. Females are a duller green and lack the elaborate tail plumes. Quetzals are found in montane and cloud forests from Panama through Guatemala and southern Mexico.

I met the resplendent quetzal at El Mirador Forest near Boqueté, Panama, in June of 1981. It rained throughout the previous night and threatened to be too muddy for our ascent into the quetzals' highland forest. The morning was dry, however, and my friends and I went off to meet the phoenix. The dry weather held, and the sun began to pierce the dense cloud cover. As we neared El Mirador, treetops were bathed in mist. The parklike forest was located on a steep ridge. Trees were moist, heavily laden with epiphytes, especially orchids and bromeliads. We looked into the canopy, searching for quetzals, only to be greeted by a troop of black howler monkeys. Soon, however, we saw our first quetzal, a female, perched upright and serene in the lower canopy. Her mate then swooped in, tail plumes streaming behind. The male was indeed "resplendent," its turquoise green tail plumes, fluffy emerald green head, and vivid scarlet breast illuminated by shafts of sunlight. When it flew, it was a burst of metallic color. As we hiked 1.5 kilometers from the quetzals' forest back to the road, we passed workers with chainsaws cutting the forest to make farmland.

MOTMOTS

Motmots (family Momotidae) consist of eight species, all Neotropical. They are most closely related to kingfishers (page 267), with which they share a similar foot structure. Both have feet on which the outermost and middle toes are fused together for almost their entire lengths. Motmots are slender birds whose back and tail colors are mixtures of green, olive, and blue with various amounts of rufous on the breast. They have a wide, black band through the eye, and some species have metallic, blue feathers at the top of their heads. They range in size from the 7-inch tody motmot (*Hylomanes momotula*) to the 18-inch rufous motmot (*Baryphthengus martii*).

Two remarkable features of motmots are a long, raquet-shaped tail (present on most but not all species) and heavily serrated bills. The tail, which in some species accounts for more than half the bird's total length, develops two extraordinarily long central feathers. As the bird preens, sections of feather barbs drop off leaving the vane exposed. The intact feather tip forms the "raquet head." One may first sight a motmot as it sits on a horizontal branch in the forest understory slowly swinging its tail back and forth like a feathered pendulum. Another distinctive motmot characteristic is its bill, which is long, heavy, and strong, with toothlike serrations. I have netted motmots in Belize and can testify as to the strength of their bite. They feed on large arthropods such as cicadas, butterflies, and spiders and will often whack their prey against a branch before eating it. They also take small snakes and lizards and frequently accompany army ant swarms (see antbirds, this chapter and chapter 2). Motmots also eat fruit, especially palm nuts, which they snip off while hovering in a manner similar to trogons.

All motmots are burrow nesters, another characteristic they share with kingfishers. They excavate a tunneled nest along watercourses or occasionally within a mammal burrow.

Motmots are vocal at dawn. The call of the common and widespread blue-crowned motmot (*Momotus momota*) may have given the family its name. The bird makes a soft and easily imitated "whoot whoot; whoot whoot." Often a pair will call back and forth to one another.

TOUCANS, ARACARIS, AND TOUCANETS

Perhaps more than any other kind of bird, toucans symbolize the American tropics. With a boat-shaped, colorful bill almost equal in length to its body, the toucan silhouette is instantly recognizable. As it flies, with neck outstretched, a toucan appears to follow its own oversized bill. *Toucan* comes from *tucano*, the name used by Topi Indians in Brazil. Altogether, there are thirty-seven species in the family Ramphastidae, including toucans, aracaris, and toucanets, and all are Neotropical. They are anatomically allied with woodpeckers and share certain characteristics of foot anatomy (two toes face foward, two face to the rear) as well as the habit of roosting and nesting in tree cavities. Toucans occur in lowland moist forests and montane cloud forests. They range in body size from 12 to 24 inches.

The huge bill is both lightweight and colorful. It is supported by bony fibers beneath the outer horny surface. The upper mandible is slightly down-curved, terminating in a sharp tip. Colorful patterns adorn most ramphastid bills.

Toucans have rather slender bodies with relatively short tails. Like their bills, their bodies are colorful, including patches of green, yellow, red, and white. One major group has ebony body feathers offset by white or yellow throats and scarlet on the rump or under the tail. Most species have a colorful patch of bare skin around each eye.

The 20-inch keel-billed toucan (*Ramphastos sulfuratus*) is one of the larger species, ranging from tropical Mexico through the upper Amazon. It is black with a bright yellow throat and breast, a white rump, and scarlet under the tail. The bill is green with an orange blaze on the side and blue on the lower mandible. The tip of the upper mandible is red, and there is bare, pale blue skin around the eye. Both male and female look alike. The keel-billed toucan has a call remarkably like a treefrog: "preep, preep, preep." Like most toucans, keel-bills associate in flocks of up to a dozen or more individuals. Typically, when one toucan flies, soon another follows, and then another. A loose "string" of toucans will move from one tree to another.

Toucans are primarily fruit-eaters, taking a wide variety of fruits from many genera including *Cecropia* and *Ficus*. They show a preference for the ripest fruits, selecting black over maroon and maroon over red, the precise order of ripest to least ripe. Toucans are relatively large, heavy birds and prefer to perch on strong branches, reaching out to snip food with their elongate bills. Trogons are able to snip off fruit at branch tips by hovering.

Toucans are gulpers (page 173). A bird snips off a fruit and holds it near the bill tip. It then flips its head back, tossing the fruit into its throat. Though this may seem awkward, the birds seem to have little difficulty. The long bill may be adaptive in permitting the relatively heavy toucan to reach out and clip fruits from branch tips, which its weight would otherwise prevent it from doing. In addition to fruits and berries, toucans eat insects, spiders, lizards, snakes, and nestling birds and eggs, all of which contain more protein than fruit.

Aracaris (genus *Pteroglossus*) are 15–16 inches long, mostly dark in color with banded breasts highlighted by bright yellow or orange red. Their bills are patterns of gray and black. They have longer pointed tails than typical toucans. Toucanets (genus *Aulachorhynchus*) are about 13 inches long and primarily greenish, with rufous tails. Their bills are dark below and yellowish above. Both aracaris and toucanets are gregarious fruit eaters.

One evolutionary note of interest about toucans is their anatomical and ecological similarity to Old World hornbills (family Bucerotidae). Both families consist of colorful birds with huge downcurved bills and slender bodies. Both nest in tree cavities, and both include fruit as a major part of an otherwise broad diet. Hornbills and toucans are not evolutionarily closely related and, thus, represent an example of convergent evolution.

Fruit and Nectar Feeders

Toucans are but one of several major bird families that concentrate on fruit for their diets. During the tropical year there is essentially a constant supply of both fruit and nectar. Though seasonal trends exist and exert important effects on animal communities (chapter 1), it is nonetheless true that some plants are fruiting or flowering every month of the year. In the temperate zone, fruits tend only to be abundant from midsummer through autumn. Many birds, including migrating species, switch over from predominantly insect to fruit diets at that time. In the tropics, however, no such dramatic switch need be made. The *constant* availability of nectar and fruit has made it possible for several major bird families to specialize and feed on one or the other (or both).

HUMMINGBIRDS

Nectar feeders consist almost exclusively of humming-birds (family Trochilidae, order Apodiformes). There are 319 species of these small, rapid fliers, all restricted to the New World. Most are tropical, but 16 species do migrate to breed in North America. The iridescent beauty of their plumage is reflected in their names: berylline, emerald-chinned, magnificent, garnet-throated, sparkling-tailed, ruby topaz, jewelfront, blossomcrown, and so on. Skutch (1973) provides a good general account of hummingbird natural history.

Hummingbirds are highly active and can easily fly foward, backward, or hover. Some of the smaller species resemble large insects as they buzz by. They accomplish their remarkably controlled flight both by a unique rotation of their wings through an angle of 180 degrees and by having an extemely high metabolism. Hummingbird heart rates reach 1260 beats per minute, and some species beat their wings approximately 80 times per second! Hummingbird metabolisms are so high that they must eat many times per day to adequately fuel their tiny bodies. Some mountain and desert species undergo nightly torpor, an adaptation to the cold temperatures of the evening. Metabolic rate and body temperature drop during the nighttime hours, and the bird is thus able to sleep without consuming an inordinate amount of energy and literally starving itself.

Hummingbirds are both thrilling and frustrating to watch because they move so quickly. Suddenly appearing at a flower, its long bill and tongue reaching deep within the blossum to sip nectar, a bird will briefly hover, move to a different flower, hover, and zoom off. Others will come and go, and some will occasionally perch. The best way to see hummingbirds well is to observe at a flowering tree or shrub with the sun to your back so that the metallic, iridescent reds, greens, and blues will glow. In those hummingbird species that are sexually dimorphic, the male has a glittering red, green, or violet-blue throat

Some hummingbirds, showing variation in bill and body size.
Rufous-crested coquette (upper right), green-fronted lancebill
(middle right), long-tailed hermit (left), purple-crowned fairy
(lower center).

patch called a gorget. The gorget is part of the male's display behavior when courting females. Depending on the sun's angle relative to the bird and observer, the gorget may appear dull, partially bright, or utterly brilliant and sparkling. When a male is courting, he positions himself such that the female is exposed to the gorget at its utter brightest.

All hummingbirds are small, the tiniest being the bee hummingbird (*Calypte helenae*) found only in Cuba and the Isle of Pines. It weighs about as much as a dime. The largest is the 9-inch giant hummingbird (*Patagona gigas*) of the Andean slopes (page 27). This bird is often first mistaken for a swift as it zooms past.

The diversity of bill anatomy, plumage, and tail characteristics among hummingbird species represents a fine example of adaptive radiation (page 133). The Andean sword-billed hummingbird (*Ensifera ensifera*), which lives high among Andean dwarf forests, has a body length of 5.2 inches plus a 4-inch-long bill! This extraordinary length is a probable case of coevolution, since the Andean swordbill feeds on *Passiflora mixta*, a flower with a very long, tubelike corolla. The booted racket-tail (*Ocreatus underwoodii*), also a cloud forest dweller, has two long central tail feathers with bare shafts but feathered at the tips, like a motmot. The ruby topaz (*Chrysolampis mosquitus*), a lowland forest and open area generalist species, is considered by many to be the most beautiful hummingbird. Males have glowing, orange throats and bright, metallic, crimson heads.

Though many hummingbirds are brilliantly colored, not all are. Among the commonest in lowland forests are the twenty-nine hermit species (genus *Phaethornis*). Most are greenish brown with grayish or rufous breasts. Unlike most species where males are brighter than females, hermits have similar sexes. All hermits have a black line bordered by white through the eyes and a long, often downcurved bill. Hermits inhabit the forest understory and edge, and their more subdued plumage seems to fit well with the dark forest interior. Male long-tailed hermits (*P.*

superciliosus) are both abundant and vocal throughout Central and South America. Males gather in courtship areas called *leks* (see manakins, this chapter) and twitter vociferously at each other as each attempts to entice a passing female.

Hummingbirds are attracted to red, orange, and yellow flowers and a single flowering tree or shrub may be a food resource for several species. When a tree is abundantly covered by flowers it is neither economical nor practical for a single hummingbird to try to defend it from others. Nonetheless, hummingbirds are generally pugnacious, and it is easy to observe both intra- and interspecific aggression among hummingbirds as they jockey for a position at their favorite flower. This competition is exacerbated because, though a plant may have many flowers, very few may be nectar-rich (see below). Some hummingbirds are highly territorial, defending a favored feeding site, while others seem to circulate along a regular route visiting several flowers. These latter are called "trapliners."

Larry L. Wolf (1975) reported an interesting example of hummingbird territorial behavior for the purple-throated carib (*Eulampis jugularis*). Males are pugnacious and territorial, defending favored flowers and dominating access to the nectar-rich food. Some females, however, employ a behavioral strategy that permits them to circumvent male dominance and gain access to desired flowers. Females court males, even during the nonbreeding season (when they cannot become pregnant). Both during and after the courtship process, a normally aggressive male will permit the "cooperative" female to feed on "his" flowers. Males inseminate females, though to no avail. Wolf reasoned that the behavior is adaptive for females, since they gain access to food that otherwise they could not hope to acquire, but he was unable to identify any clear advantage to the male, since courtship and copulation use energy and no offspring result. Wolf titled his paper " 'Prostitution' Behavior in a Tropical Hummingbird."

Hummingbirds often have a mutualistic relationship with plants, feeding on nectar but facilitating cross-pollination. Hermits, for instance, often feed on the nectar of heliconia flowers (chapter 2). Many heliconias produce relatively constant amounts of nectar per flower. However, one heliconia studied in Costa Rica by Peter Feinsinger (1983), *Heliconia psittacorum*, exhibits what Feinsinger calls a "bonanza-blank" pattern of nectar production. Some flowers contain abundant nectar (bonanzas), some essentially none (blanks). Many other tropical plants, especially those in open successional areas, also are bonanza-blank flowerers (see below). Hermits must visit many flowers in order to encounter one with high nectar content, thus the bonanza-blank pattern presumably aids *Heliconia psittacorum* in accomplishing cross-pollination.

In a comprehensive study of ten successional plant species and fourteen hummingbird species at Monteverde, Costa Rica, Peter Feinsinger (1978) documented that flowering was staggered among plant species, resulting in a constant nectar supply to hummingbirds. In five plant species that were closely measured for nectar volume, the bonanza pattern was highly evident. Feinsinger speculated that plants may conserve energy by producing large numbers of "cheap" nectarless flowers and a mere few "expensive" bonanza flowers, forcing hummingbirds to visit many flowers to find satiation. By visiting many flowers, cross pollination is promoted.

Hummingbirds display a range of foraging patterns. Peter Feinsinger and Robert K. Colwell (1978) identified six patterns evident in how hummingbirds exploit flowers: high-reward trapliners, which visit but do not defend nectar-rich flowers with long corollas; territorialists, which defend dense clumps of somewhat shorter flowers; low-reward trapliners, which forage among a variety of dispersed or nectar-poor flowers; territory-parasites of two types (large marauders and small filchers); and generalists, which follow shifting foraging patterns among various resources. Large marauders are species with large bodies that can intimidate normally territorial smaller species. They move in and take what they want. Small filchers are species with small bodies that "sneak" in to feed quickly, before being detected by a territorial bird. High-reward trapliners such as the hermits have a regular route that they visit and are most common in the forest understory. The other types of foraging are evident in the canopy and open, successional areas.

The complexity of interactions between hummingbirds and plants was well elaborated by research conducted by Robert K. Colwell (1973, 1985a, 1985b) that revealed how nectar-eating mites (*Rhinoseius*) depend entirely upon hummingbirds for their dispersal among flowers. The mites are transported in the nasal cavities of the birds! Colwell showed that mites were dependent not on either the birds or plants alone, but on the *mutualistic* interaction between birds and plants. Colwell unraveled a complex ecological interdependence among two mite species, three hummingbirds, one flowerpiercer (next paragraph), and four hummingbird-pollinated plants.

One other group of birds, besides hummingbirds, feeds principally on nectar. The eleven species of flowerpiercers, all members of the large tanager family (see below), do not probe into the center of the flower, as hummingbirds do but instead snip a minute hole through the petal at the base. The bird pokes its bill in, sips nectar, but receives no pollen. Flowerpeircers are therefore nectar parasites. A few hummingbirds have occasionally been seen employing a similar behavior.

TANAGERS

Tanagers are a group of very colorful, small perching birds. *Tanager*, like the word *toucan*, comes from the Topi Indian language of Brazil. The diverse subfamily Thraupinae (part of the huge family Emberizidae, order Passeriformes) consists of 242 species of tanagers, euphonias, chlorophonias, honeycreepers, dacnis, concbills, and flowerpiercers. Most species are brilliantly colored, and feed on fruit (mashers, page 173), nectar, and insects. All occur in the New World, and they are found abundantly from lowland forests to high montane and cloud forests. They are particularly common around forest edge habitats and are easy to see at fruiting figs, cecropias, etc. Though 4 tanager species migrate to North America to breed, the remainder are all confined to the Neotropics. Storer (1969) reviews a general taxomony of tanagers, and Isler and Isler (1987) provide a comprehensive field guide, illustrating all species.

In most Neotropical tanagers, males and females have similar plumage. The common names of tanager species reflect their multicolored, exotic feather patterns. On a trip to Panama, for instance, one may encounter the crimson-collared, scarlet-rumped, flame-colored, blue and gold, golden-hooded, silver-throated, and emerald tanagers, a list that is far from exhaustive. One of the most common and widely distributed birds of the tropics is the blue-gray tanager (*Thraupis episcopus*), which is well described by its name. Chlorophonias are bright green and yellow, highland tanagers. Among the most exotically colored tanagers is the paradise tanager (*Tangara chilensis*) of South America. Like a mosaic of neon colors, this incredible bird has a golden green head, purple throat, bright scarlet lower back and rump, black upper back, and turquoise breast. Euphonias (genus *Euphonia*) comprise a group of small tanagers, also quite multicolored, that tend to feed heavily on mistletoe berries (family Loranthaceae). They are important seed dispersers of mistletoe, as their sticky droppings, deposited on branches, contain the seeds that begin life as epiphytes, before becoming parasitic. Euphonias nest in bromeliads (Bromeliaceae).

Honeycreepers, which include dacnises and conebills, are nectivorous, though they also include ample amounts of fruit and arthropods in their diets. Warbler-sized, they have fairly long, downcurved bills. One of the most common and widely distributed of this group is the bananaquit (*Coereba flaveola*). This small bird with a dark back, white eye stripe, and yellow breast is among the most ubiquitous of tropical birds. It is found in virtually all habitats from lowlands to cloud forests, and often becomes quite tame around gardens. Like the flowerpiercers, bananaquits are prone to poke a small hole at the outside base of a flower and drink nectar without contacting pollen.

Some tanagers, such as the ant-tanagers (genus *Habia*) are army ant followers, and many tanagers, euphonias, and honeycreepers move with antbirds, woodcreepers, and other species in large mixed foraging flocks. Charles A. Munn (1984, 1985) and John Terborgh (Munn and Terborgh 1979) have studied species compositions of both canopy and understory mixed species flocks in the Peruvian Amazon. Each flock type consists of a core of five to ten different species, each represented by a single bird, a mated pair, or a family group. Up to eighty other species join flocks from time to time, including twenty-three tanager, euphonia, and honeycreeper species, a remarkably high diversity. Mixed foraging flocks occupy specific territories and, when another flock is encountered, the same species from each flock engage in "singing bouts" and displays as boundary lines are established. Adult birds tend to remain flock members for at least two years. Nesting occurs in the general territory of the flock, the nesting pair commuting back and forth from nest to flock.

Munn's work revealed an odd twist on interactions within mixed foraging flocks. One long held hypothesis about mixed flocks is that being part of a mixed flock serves to help protect against predation (Moynihan 1962). With so many eyes looking, predators have difficulty going undetected. Munn's observations showed that sentinel species are part of the mixed flock. One species of shrike-tanager and one antshrike (an antbird, see below) aided the flock by giving alarm calls when danger threatened. Both sentinel species, however, also gave "false alarm calls," a behavior Munn described as "deceitful." False alarm calls are exactly that: alarm calls uttered when *no* danger is present. The hypothesized reason for the false alarms calls is that the alarmist has a better chance of capturing food that is also being sought by another bird. When a white-winged shrike-tanager (*Lanio versicolor*) and another species were both chasing the same insect, the shrike-tanager's false alarm would cause momentary hesitation by the other bird, allowing the shrike-tanager to capture the insect.

ORIOLES, OROPENDULAS, AND CACIQUES

The large avian family Icteridae (order Passeriformes) includes the blackbirds and their relatives. Among those occurring in the Neotropics are thirteen oropendula species, nine cacique species, and twenty-four oriole species. Oropendulas and caciques are colonial and make long, hanging, basketlike nests. An oropendula nest tree is easy to spot because it is out in the open and adorned with numerous pendulous nests. The isolation of the nest tree affords some protection against predation by monkeys, since the simians are usually loathe to leave the canopy and cross open ground. Oropendulas are large birds (some almost crow-sized), and caciques are robin-sized. In shape, both caciques and oropendulas are relatively slender with long tail and sharply pointed bills. Oropendulas come in two color types. One group of species is mostly black and chestnut, with yellow on the bill and tail, and the other is quite greenish. Caciques are mostly sleek black but with bright red or yellow rumps and/or wing patches, and yellow bills.

Both caciques and oropendulas tend to locate their colonies near bee or wasp nests. Because these colonial insects can be very aggressive toward intruders, this behavior helps reduce the probability of predation by mammals. Scott K. Robinson (1985a, 1986), who studied the yellow-rumped cacique (*Cacicus cela*) in the Peruvian Amazon, learned that caciques employ other "strategies" that would also seem to protect the colony. Caciques often nest on islands in a river or lake, affording added security from both mammals and snakes, for would-be predators would have to cross a water body patrolled by otters and caymans. Robinson also noted that caciques tend to mob potential avian predators and that they permit unused abandoned nests to remain in the nest tree along with active nests. The presence of the unused nests presumably confuses a predator. Not surprisingly, Robinson found that each cacique attempts to locate its nest in the center (where protection is maximized), rather than the more risky periphery of the colony. Robinson's overall conclusion was that colonial nesting and group defense is a significant adaptation against nest predation.

Robinson documented, however, that yellow-rumped caciques were occasional victims of nest piracy by other bird species. One, appropriately named the piratic fly-catcher (*Legatus leucophaius*), harassed caciques until they abandoned their nests to the flycatchers. Russet-backed oropendolas (*Psarocolius angustifrons*) destroyed cacique eggs and killed young, leaving empty nests. Finally, troupials (*Icterus icterus*), which are large, aggressive orioles, both took over cacique nests and destroyed eggs and young. Robinson hypothesised that the piracy is not related to competition for food because each of the nest pirate species has a diet different from that of caciques. Instead, Robinson argues that the creation of many nearby empty cacique nests, which he describes as a "maze," serves to confuse potential predators and confer protection on the nest pirate species (Robinson 1985b).

Orioles nest as territorial pairs and are not colonial like the oropendulas or caciques. They are colorful, with various combinations of orange, yellow, and black. Several oriole species migrate to nest in North America, but most remain in the tropics.

Orioles, oropendulas, and caciques feed heavily but not exclusively on fruit and nectar, mixing various arthropods into an otherwise vegetarian diet. Like the tanagers, these birds move in mixed flocks foraging in fruit trees and drinking nectar from blossoms. In Belize, I have seen flocks of wintering orchard (*Icterus spurius*) and northern (*I. galbula*) orioles, visiting cecropias along with summer tanagers (*Piranga rubra*) and various Neotropical honey-creepers, orioles, and tanagers.

PARROTS

Like toucans, parrots are quintessentially tropical. Global in distribution, they occur mainly in tropical forests of the southern hemisphere. In the Neotropics, 136 species of the family Psittacidae (order Psittaciformes) can be found ranging from the spectacular large macaws (genus *Ara*) to the sparrow-sized parrotlets (genus *Forpus*). Among the most commonly encountered of the New World parrots are the chunky short-tailed Amazons (genus *Amazonia*). Parrots are mostly green (though there are some dramatic exceptions) and can be remarkably invisible when perched in the leafy forest canopy. They reveal their presence by vocalizing, usually a harsh screech or squawk. Often a flock will burst from a tree like shrieking banshees, and it is amazing to see how many birds were actually in the tree when so few were readily visible. Generally, there is no difference in plumage between the sexes. Forshaw (1973) has treated all of the world's parrots, including, of course, all Neotropical species.

Parrots are gregarious frugivores. It is uncommon to find only one or two. Flocks move about in forests and savannas searching out fruits, flowers, and occasionally roots and tubers. Parrots climb methodically around the tree branches, often hanging in awkward acrobatic positions as they attack their desired fruits. Their sharply hooked, hinged upper mandible is useful in climbing around in trees as well as in scraping and scooping out large fruits. Using their strong nutcrackerlike bills, they can crack many of the toughest nuts and seeds, which they eat with equal relish as the pulpy fruit itself. Their tongues are muscular, and they are adept at scooping out pulp from fruit and nectar from flowers. Because of their ability to crush and digest seeds, they are primarily seed predators rather than seed dispersers (page 174). Daniel Janzen (1983c) studied the orange-chinned parakeet (*Brotogeris jugularis*) in Costa Rica and learned that it acted as a seed predator on the trees *Bombacopsis quinatum* and *Ficus ovalis*. Janzen shot one bird, examined its crop, and found several thousand *Ficus* seeds all of which were damaged in some way. Droppings from orange-chins contained no evidence of intact seeds.

The most spectacular Neotropical parrots are the macaws. Large, long-tailed parrots, macaws range in plumage from the predominantly green chestnut-fronted (*Ara severa*), military (*A. militaris*), and great green (*A. ambigua*), to the bright red scarlet (*A. macao*), and red-and-green (*A. chloroptera*), to blue-and-yellow (*A. ararauna*), to the deep blue of the hyacinthine (*A. hyacinthinus*). Macaws are most commonly seen flying to and from their roosting and feeding sites. Their slow wing beats and long tails make them very distinctive in flight. Many macaws frequent gal-

Scarlet macaw.

lery forests along watercourses or humid forests interrupted by open areas.

COTINGAS

Cotingas (family Cotingidae, order Passeriformes) are
among the real glamour birds of the Neotropics. With
names such as bellbirds, umbrellabirds, cocks-of-the-rock,
pihas, fruiteaters, fruit-crows, and purpletufts, the sixty-
five cotinga species (Snow 1982) comprise a colorful and
diverse family, most of which are confined to the lowland
tropical forests. They are birds of rain forests and jungles
and are described by David Snow (1982) as "extreme fruit
specialists." Large cotingids eat laurels (Lauraceae), incense (Burseraceae), and palms (Palmae), while smaller
species eat smaller, sweeter fruits, often plucking fruits
while hovering. Cotingas typically have wide, flattened
bills, shaped well for accommodating rounded fruits. Cotingas feed only on the flesh of the fruit and not the seeds
and thus can be effective seed dispersal agents. Some species, such as the fruit-crows and pihas, mix insects among
their fruits, but most cotingas feed exclusively on fruit.

Cotingas are diverse. Some, such as the umbrellabirds
and cocks-of-the-rock, are large and colorful or have ornate plumage, while others, such as the fruiteaters and
pihas are smaller and relatively drab. Some are sexually
monomorphic, the males and females looking alike, while
others represent extreme cases of sexual dimorphism.
Some form pairs and occupy territories, while others are
highly polygynous, cocks mating with many hens. In a
few species, such as the cocks-of-the-rock (genus *Rupicola*)
and screaming piha (*Lipaugus vociferans*), males gather to
court females in mating areas called leks. Bellbirds (genus
Procnias) are known for their piercing bell-like call notes,
pihas for their loud "scream," cotingas for their shiny metallic plumage, cocks-of-the-rock for their golden-orange
or orange-red coloration and fan of head feathers, and
umbrellabirds (genus *Cephalopterus*) for their extraordinary umbrellalike head plumes and inflatable scarlet air
sac on the breast. Cotingas generally make small incon-

spicuous nests, incubate but a single egg, and have a prolonged incubation period. Bellbirds typically incubate for approximately thirty days, and cocks-of-the-rock for forty or more days. This long incubation period is probably related to feeding nestlings almost exclusively fruit, which is low in protein but high in fat and carbohydrate (see below).

MANAKINS

The fifty-nine species of the family Pipridae (order Passeriformes), the manakins, are small, chunky fruit-eating birds most of which inhabit lowland forests. Males of most species are quite colorful; females, drab olive green and yellowish. Manakins have short tails, rounded wings, and a short but wide bill with a small hooked tip. They pluck fruits on the wing, supplementing their fruit diets with arthropods. Manakins have among the most elaborate courtship displays of any birds. Many species court in leks, assemblages of males that display to transient females. In addition to being brightly colorful, male manakins "dance" in elaborate fashion before females. Manakin courtship is detailed in the next section. Only females build the nest, incubate the egg, and feed the young. Clutch sizes are typically quite small, one to two

Male white-bearded manakin.

birds per nest. Manakin courtship is reviewed by Sick (1967), Lill (1974), and Snow (1976).

Leks and Lovers—Sexual Selection among Cotingas and Manakins

Charles Darwin (1859, 1871) devised his theory of sexual selection in part to account for why certain bird species, among them many of the cotingas and most manakins, display extreme differences in plumage between the sexes. This sexual dimorphism almost always involves brightly and ornately plumaged males compared with subtle-plumaged, more cryptically-colored females. Why females are cryptic seemed an easy question to Darwin. Females undergo natural selection for cryptic plumage because such coloration aids in reducing the risk of discovery by predators. But why are males colorful? Adding to this mystery was the fact that elaborately colored males often augment their already gaudy selves by engaging in bizarre courtship displays.

THE AMOROUS COCK-OF-THE-ROCK

The Guianan cock-of-the-rock (*Rupicola rupicola*), a large cotinga, provides an example of elaborate courtship and plumage. This species has been studied by Snow (1982) and Trail (1985a, 1985b). Males are chunky, with short tails and bright golden orange plumage but black on the tail and wings. Beaks, legs, eyes, and even the very skin is orange! This already striking plumage is further enhanced by delicate elongate orange wing plumes and a crescent like thick fan of feathers extending from the base of the bill to the back of the neck. Females are dull brown, with neither the wing plumes nor the head fan. Males gather in courtship areas called leks within the rain forest. Each male clears an area of ground in which to display and defends perches in the vicinity of its display site. The lek can be a crowded place, with males as close to one another as 4–5 feet and several dozen males on the same lek. When a female approaches a lek, each male dis-

Male Guianan cock-of-the-rock.

plays by landing on the ground and posturing to her. Each displaying cock strokes its wing plumes and turns its head fan sideways so that the female sees it in profile, and stares at her with its intense orange eye set against flaming orange feathers. The object of the cock's bizarre display is to mate, presumably by suitably impressing the female. Females do not appear to be easily impressed. A hen will typically visit a lek several times before engaging in copulation. These visits, called mating bouts, always excite the males to display. Only one male on the lek will get to mate with a visiting female, who may return to mate with him a second time before laying eggs (Trail 1985b). No extended pair bond is formed, only a brief coupling. The cock returns to the lek, continuing to court passing hens while the newly fertilized hen attends to nest building, egg laying, incubation, and raising the young. The basis of her behavior in choosing a male from among many potential contenders is one facet of what Darwin called *sexual selection*.

Darwin reasoned that, in some species, female choice was the dominant factor in selecting males' appearances. Put very simply, males are pretty (or musical or noisy or perform complex "dances") because females have tended through generations to mate mainly with males having these unique features. Since plumage color is heritable (as are behavioral rituals), gaudy coloration was selected for and continually enhanced.

The other facet of sexual selection recognized by Darwin is that males must compete among themselves for access to females. This may be accomplished by dominance behavior, guarding females, active interference with other males' attempts to mate (see below), injury to other males, or merely being "sneaky," and mating before other males can react. Gaudy plumage may contribute to a male's success by intimidating other males and thus make it easier to gain the attentions of a female. Male/male competition coupled with female selection of the "winner" is what Darwin defined as sexual selection.

Sexual selection has costs for both males and females. Though the hen exercises the most choice in the mating process, she is left solely responsible for the chores of nest building, incubation, and caring for the young. Males may at first glance seem the luckiest, rewarded by a life of lust in nature's tropical "singles bar," the lek. The combination of male/male competition plus dependency on female choice makes life difficult for most males, however. Though some cocks are quite successful, mating frequently, others, the losers, spend their entire lives displaying to no avail. After a lifetime of "frustration," they die genetic losers, never selected even once by a hen. Pepper W. Trail (1985a, 1985b), who studied the Guianan cock-of-the-rock in Suriname, documented high variability in male mating success. He found that 67% of territorial males failed to mate at all during an entire year. The most successful male performed an average of 30% of the total number of annual matings, and the lek contained an

average of fifty-five cock birds! One of these fifty-five
mated 30% of the time. Many never mated. Such is the
cost of sexual selection for males. In reproductive terms,
females are the most fortunate sex. Most females do
mate, though success in fledging young may vary among
females.

Pepper Trail (1985a) also discovered another interesting twist in the mating process of the Guianan cock-of-
the-rock. Some males were sore losers and habitually disrupted the mating of others. Trail found that aggressive
males that disrupted copulations by other males fared
better in subsequent mating attempts. He learned that
males that were confrontational "were significantly more
likely to mate with females that they disrupted than were
non-confrontational males." He hypothesized that only
the cost of confrontation in terms of energy expenditure,
loss of time from the aggressive bird's own lek territory,
plus risk of actual retaliation kept direct confrontational
behavior from becoming even more manifest among the
birds. On the other hand, Trail (1985b) found adult fully
plumaged males remarkably tolerant of juvenile males
that were still plumaged in drab colors, resembling females. Yearling males would actually attempt to mount
adult males as well as females in a crude attempt at mating. Adult males did not respond aggressively to these
misguided efforts, possibly because yearling plumage,
being drab, does not stimulate an aggressive response.

SCREAMING PIHAS AND CLANGING BELLBIRDS

Sexual selection has evolved in various ways, and thus
courtship patterns differ among species. The screaming
piha (*Lipaugus vociferans*), like the cock-of-the-rock, courts
on leks, assemblages of males each soliciting females. The
piha differs from the cock-of-the-rock in that it is not
elaborately plumaged. In fact, it is downright nondescript, being a slender, robin-sized bird light gray on the
face and breast and dark gray-brown on wings, back, and
tail. Though certainly attractive (at least to human eyes),
it is by no means the glamour bird that its cousin is. In

fact, male and female screaming pihas look alike and, thus, would not seem to fit with Darwin's concept of sexual selection, focused as it is on sexual dimorphism. It is with voice, however, and not looks, that a male screaming piha attracts a female. Piha leks are comprised of up to thirty males, and David Snow (1982) reports that males' calls are "one of the most distinctive sounds of the forests where these birds occur, its ringing, somewhat ventriloquial quality seeming to lure the traveller ever onwards into the woods." Snow describes the call as "pi-pi-yo" or "cri-cri-o." Barbara Snow studied the screaming piha in Guyana and found that one male spent 77% of his time calling on the lek, usually from a thin horizontal branch well below the canopy. An excited male called at the rate of twelve times per minute. Calling seems to replace plumage and display behavior as the signal to the females. Sexual selection has occurred, but for characteristics of voice, not appearance.

Bellbirds (genus *Procnias*), like the screaming piha, rely heavily on voice as part of the courtship process. There are four species, each shaped generally like a starling, though larger in size, ranging throughout lush montane forests of northern South and Central America. They tend to migrate vertically, breeding in highland forests and moving downslope to lowland forests when not breeding. Unlike pihas, bellbirds are sexually dimorphic, the males having much white on the body along with ornate wattles about the head. In one species, the white bellbird (*P. alba*), the male is entirely white with a fleshy wormlike wattle dangling from its face above the bill. The male bare-throated bellbird (*P. nudicollis*) is almost all white but has bare blue skin on the throat and face around the eyes. The male bearded bellbird (*P. averano*) has black wings and a chestnut head with a heavy "beard" of black fleshy wattles hanging from its throat, and the male three-wattled bellbird (*P. tricarunculata*) is chestnut on body, tail, and wings, but with a white head and neck, and three fleshy wattles hanging from the base of the bill.

Male three-wattled bellbird (upper) and
male bearded bellbird (lower).

Females of all four species are similar greenish yellow,
darkest on the head, with streaked breasts.

Male bellbirds establish calling and mating territories in
the forest understory. Though not true lek birds, bellbird
courtship territories are closely spaced together. Each
male spends most of his time on territory vocalizing to
attract hens. Males take no part in nest building, incubation, or raising young. I have observed the bearded bellbird in Trinidad and the three-wattled bellbird in Panama. Both court in a generally similar manner.

Male bearded bellbirds are among the first sounds one

hears upon entering the Arima Valley in Trinidad. David Snow (1976, 1982) aptly describes their call as a loud "Bock!" The call carries amazingly well, and I thought the birds were nearby when, in reality, they were a quarter of a mile or more from me. The call definitely has a bell-like quality, though it is a muted clang, and the ventriloquial quality of the call note is evident. Even when very close to a calling male, it can be frustratingly difficult to locate him. Cock birds initially call from a perch above the canopy, often on a dead limb, but will drop down into the understory to complete the courtship. Females never call, and it is clear that male vocalizations are an essential part of sexual selection in bellbirds.

The object of calling is to attract a hen to the male's territory. Each cock bellbird has his own courtship site in the forest understory. The male "bocks" rather continuously, mixing the bocking with a series of "tock, tock, tock" notes. If successful in luring a female to his territory, the male initiates a series of courtship postures, performed from a horizontal branch upon which the female perches as his only audience. These postures include display of the beard wattles, a wing display, and a display in which a bare patch of skin on the male's thigh is revealed. All bellbird species include a "jump-display" as part of courtship. A cock bearded bellbird will leap from one perch to another, landing before the hen with his body crouched, tail spread, and eyes staring at her. You can guess what happens next, assuming the male has performed satisfactorily.

The three-wattled bellbird takes the jump-display one step further. The cock jumps over to the place occupied by the hen while at the same moment the hen skitters along the limb to occupy the place the male just vacated. Called a "changing-place" display, the male then slides across the branch to be right next to the hen, emitting a "close-up" call virtually in her ear. Following the successful execution of this maneuver, more bellbirds come into the world.

THE DANCING MANAKINS

Manakins carry the fine art of courtship dancing to extremes. Male manakins are brightly colored, glossy black with bright yellow, orange-red, or golden heads and/or throats, some with bright yellow or scarlet thigh feathers, and some with deep blue on their breasts and/or backs and long streamerlike tails. A few species are sharply patterned in black and white. But fancy feathering notwithstanding, it is dancing in which these birds excel. David Snow (1962a, 1976), Helmut Sick (1967), and Paul Schwartz and David Snow (1978) have made detailed studies of manakin courtship behavior and have provided much of the information outlined below.

The white-bearded manakin (*Manacus manacus*) courts on rain forest leks. I've observed its courtship antics in the Arima Valley in Trinidad. The male has a black head, back, tail, and wings but is white on the throat, neck, and breast. Its name comes from its throat feathers, which are puffed outward during courtship. Females are greenish yellow. Up to thirty or more males may occupy a single lek. Each male makes his "court" by clearing an oval-shaped area of forest floor 2 to 3 feet across. Each court must contain two or more thin vertical saplings, as these are crucial in the manakin's courtship dance. The male begins courtship by jumping back and forth between the two saplings, making a loud "snap" with each jump. The snap comes from modified wing feathers snapped together when the wings are raised. When a female visits the lek, the snapping of many males is audible for quite a distance. In addition to the snap, the male's short wing feathers make a buzzing insectlike sound when it flies, and thus active manakin leks can become a cacophony of buzzing, snapping birds. The intensity of the male's jumping between saplings increases until he suddenly jumps from sapling to ground, appearing to ricochet back to another sapling, from which he slides vertically downward like a fireman on a pole. David Snow's film footage of the slide revealed that successful males slide down

Courtship dance of the male white-bearded manakin. See text
for details. From H. Sick (1967). Reproduced with permission.

right to the female perched at the base of the sapling
"pole." Copulation is so quick that Snow only discovered
the presence of the female in the film. He never saw her
while he was witnessing the event!

Following copulation, the female leaves the lek and attends to nesting. The male starts to dance again. Male
manakins spend most of their adult lives at the lek. Some,
as in the case of the cock-of-the-rock, are probably highly
successful and mate very often. Others may never mate.
Observations of banded males on Trinidad have revealed
that life on a lek is usually fairly long for individual birds.
Some live for a dozen years or more, a very long life span
for such a small bird (Snow 1976). Males generally only
leave the lek to feed on ripe fruits.

Another common Trinidad manakin that I have observed, the golden-headed (*Pipra erythrocephala*), is not a lek dancer, but rather each male displays in his own territory. As in the white-bearded, the dance begins when the male darts back and forth on selected twigs, calling "zlit" as he does so. Unlike the white-bearded, which dances close to the ground, the golden-headed usually displays about 10 feet off the ground in an understory tree. The cock becomes increasingly vigorous in his dancing, crouching, his body at a 45 degree angle as he slides along a horizontal twig. His sparkling golden head and sleek black plumage are displayed very conspicously, but more is yet to come. When a female arrives, the male skitters along the branch toward her, but *tail first!* As he advances, he bows, spreads his wings, and exposes bright red thigh feathers, all the while pivoting his body back and forth. The climax of the dance comes when the male suddenly flies from the dance branch and quickly returns, inscribing an "S-shaped" curve as he lands with wings upraised before the female. Various vocalizations accompany the performance.

If the white-bearded and golden-headed manakin performances amaze you, be warned that the blue-backed and swallow-tailed manakins (genus *Chiroxiphia*) seem to carry bird dancing to the point of incredulity. Blue-back males have bright red on the top of their heads, shimmering torquoise blue backs on otherwise shiny black plumage. Swallow-tailed manakins, also called blue manakins, are similar but with blue both on back and breast and elongate central tail feathers. Even for male manakins they are extraordinarily beautiful, especially when seen in a burst of full sunlight as they dance in the forest understory. I say *they* because these manakins dance as a team. Two blue-backed males engage in a coordinated jump dance in which both birds occupy a horizontal thin branch, one jumping and hovering while the other crouches on the branch, the other jumping and hovering when the first lands. As they dance, they vocalize. The dance may occur in the presence or absence of a female, the males seeming to "practice" when a female is not present. The dance ends when one of the three cocks bows before the hen, head turned exposing the bright red top, blue back upraised. In the case of the swallow-tailed or blue manakin, up to three males dance in perfect coordination before a single female. The three dancers align themselves horizontally on a thin branch, shoulder to shoulder before the female, each male facing in the same direction. The male farthest from the female jumps up, inscribes a 180 degree angle and lands nearest the female, next to the other males. He immediately turns around, so once again all three dancers face the same direction. A second dancer, again the farthest from the female, repeats the first dancer's performance, and so on. The dance happens rapidly, and David Snow has described it as a spinning "Catherine wheel" of dancing males, jumping, displaying, and vocalizing in total coordination. No other case of such elaborate team dancing is known for birds. The termination of the performance occurs when one of the males vocalizes sharply, the effect of which is to "turn off" the other two males. The dominant male then erects his red head feathers as he perches before the female. She and he fly off into the underbrush.

One species, the wire-tailed manakin (*Pipra filicauda*), adds yet another element to the roster of manakin courtship techniques. Males, which are black with yellow breasts and a red cap, have stiff tail feathers that terminate in long, delicate filaments. Wire-tailed males dance in teams of two, rather like the blue-backed species. However, when the dominant male approaches a female, he performs a "twist" display in which he rotates his posterior side to side, gently touching the female on her chin with his tail filaments. Females apparently respond well to this maneuver, for a female will typically slide toward a male to receive the tail brushing. This is the only known example of tactile stimulation among manakins, and it appears that the unique tail is the product of sexual selection (Schwartz and Snow 1978).

Why do several male manakins cooperate in courting a single female? Only one will get to mate with her. Some evolutionary theorists believed that the males were perhaps brothers, sharing the majority of each others' genes. Cooperative behavior could result in reproduction of many of one's own genes, that happen to be shared with one's brother, another example of possible kin selection (page 203). However, Mercedes S. Foster (1977), who studied *Chiroxiphia linearis*, found little evidence for a close relationship among cooperating males. Foster found that one male was consistently dominant over the others. Though the assemblage remained together throughout the breeding season and even from one year to the next, cooperating males were not brothers and did not behave altruistically toward one another. Rather, subordinates were biding their time until they could replace the dominant male. Foster hypothesized that one male, acting alone, could never succeed in attracting a female. Only by being part of a pair or trio could a male hope to eventually succeed. When a dominant male dies, it is quickly replaced by a subordinate who "trained" under it.

WHY DO PIHAS SCREAM, BELLBIRDS CLANG, AND MANAKINS DANCE?

The bizarre results of sexual selection in cocks-of-the-rock, pihas, bellbirds, and manakins are evident, but what sorts of selection pressures were responsible for their evolution? Both David Snow (1976) and Alan Lill (1974) have suggested possible scenarios for the "release" of males from postcopulatory reproductive chores, thus initiating the male/male competition and pattern of female choice that resulted both in the gaudy plumages and elaborate courtship behaviors.

David Snow emphasizes the importance of a diet almost exclusively of fruit. He points out that both bellbirds and manakins feed so heavily on fruit that they are easily able to secure adequate daily calories with only a small percentage of their time devoted to feeding. Fruit is both relatively abundant and easily collected. It does not have to be stalked or captured and subdued. The male bellbird or manakin has lots of time in which to clang or dance.

Alan Lill, who studied manakins, agrees that a fruit diet is significant in the evolution of sexual selection in these birds. He places his emphasis, however, a bit more on nest predation. A largely frugivorous diet has metabolic costs as well as benefits (Morton 1973). Incubation time is relatively long and nestling growth rates slow in highly frugivorous birds because fruit is nutritionally not well balanced for a baby bird (low in protein but high in fat and carbohydrate). Lill argues that because of the slow development time brought about by a diet of fruit (recall oilbird, page 175), nest secrecy is of paramount importance. Heavy egg and nestling predation are best minimized by having only one bird, the cryptically colored female, attend the nest. A male's presence at the nest could actually be detrimental to raising young, since one bird can easily find sufficient food for the small brood (usually two nestlings), and a second bird might inadvertently reveal the presence of the nest to potential predators. Lill argues that it is to the advantage of both female and male for the male to stay away because male absence actually increases the probability of egg and nestling survival. Males are dispensable, not needed for raising young. Lill concludes that this "male liberation" was followed by sexual selection and male "chauvinism" in the odd and varied forms described above.

WHY LEKS?

Given that a combination of factors have "released" males from attending nests, why have some species organized their courtship bouts in leks? Several hypotheses have been suggested. One, called the "female preference model," argues that females "prefer" groups of males when making their selections of whom to mate with (Bradbury 1981). A male that stayed away from the lek would not attract any female, thus males have no choice but to join a lek. Another suggestion is that males might associate in leks because the lek area happens to be a

place where females, for whatever reason, frequently occur. This idea, termed the *hotspot model*, presumes that
leks form rather accidently, as males gather where they
are most likely to encounter females (Emlen and Oring
1977, Bradbury and Gibson 1983). Both hypotheses place
strong emphasis on female choice as causal to lek formation.

Bruce M. Beehler and Mercedes S. Foster (1988) have
critiqued both the female preference and hotspot models
and have concluded that neither is sufficient to account
for the evolution of lek mating systems. They offer yet
another model, dubbed the *hotshot model*, that emphasizes
the role of male-male dominance and interactions between dominant and subordinate males on a lek. Hotshots are individuals that control leks. Subordinates occasionally benefit from disrupting leks (recall Trail's
observations of subordinate cocks-of-the-rock cited
above), but mostly they bide their times while slowly advancing toward dominance. Beehler and Trail argue that
novice males have little choice but to begin as subordinates, working their way up through the ranks to attain
dominance status before they can reproduce. Subordinate birds congregate around the dominant cocks, since
they have no hope for mating otherwise (recall Foster's
observations on manakins cited above). The hotshot
model places extreme emphasis on male-male interactions rather than male appearance and female choice.
Dominance among cocks can be subtle, but it is real, and
females will almost always select a dominant male with
whom to mate. Beehler and Foster offer several predictions from their model. For instance, if all hotshot (dominant) cocks are removed from a lek, Beehler and Foster
predict that disruptions among the remaining males will
increase (because none are dominant) and that the lek
may break up into several smaller leks, as new dominance
rankings are established. Removal of the hotshots also
predicts that females will visit the lek less and mate less
until the lek restabilizes. These predictions are testable
and both the female preference and hotspot models pre-

dict different outcomes. Beehler and Foster cite some evidence favoring their model, though many more tests need to be done before the hotshot model can be regarded as "dominant."

Insect-Arthropod Feeders

Several major groups of tropical passerine (order Passeriformes) birds utilize insects and other arthropods as virtually their only food sources. These groups are among the most species-rich found anywhere. For instance, there are 217 species of ovenbirds (Furnariidae), 231 species of antbirds (Formicariidae), 48 species of woodcreepers (Dendrocolaptidae), and an astonishing 367 species of tyrant flycatchers (Tyrannidae) (Austin 1985). Of the above groups, only a few of the tyrannids venture to North America to nest. All others are entirely Neotropical.

Tyrannids represent a notable case of species diversification and adaptive radiation (chapter 4, page 141). Their insect diets have probably provided major impetus in producing such diversity over evolutionary time. Eating insects *per se* does not cause species diversity nor speciation. It does, however, promote specialization, which produces divergence and can, therefore, be a factor in speciation. Insects require catching; they do not seek predators, but, on the contrary, are well adapted to avoid predation through either cryptic or warning coloration or escape behavior. Each insect-eating bird tends to develop a particular pattern of feeding, and its size, behavior, and bill shape become refined to focus on a particular size range and type of prey (Fitzpatrick 1980a, 1980b, 1985). Prey characteristics provide major selection pressures in shaping evolution among avian predators. Secondly, species compete against each other. The presence of many other insect-eating species could generate continuous diffuse competition within a species assemblage, keeping each species ecologically focused on doing what it alone does best.

Insect eaters can roughly be categorized by overall feeding method. There are (1) flycatchers (tyrant flycatchers and nunbirds), (2) bark probers and drillers (woodcreepers and woodpeckers), (3) foliage gleaners (ovenbirds), and (4) ant followers (antbirds).

FLYCATCHERS

Tyrant flycatchers have been discussed (chapter 4, page 141). There is, however, another group, less diverse but deserving of mention here. The puffbirds and nunbirds (family Bucconidae, order Passeriformes) consist of thirty-one species, all Neotropical, that feed on insects and spiders captured by darting from a perch and snatching them in midair. The black-fronted nunbird (*Monasa nigrifrons*) is typical of the group. Ranging throughout the Amazon Basin, this ubiquitous, robin-sized, forest-dwelling bird is easily recognized by its black upper plumage and tail, gray breast, and tapered, slightly drooping, bright red-orange bill. It perches upright on a horizontal limb, and, typical of "sit and wait" predators, hardly moves a muscle until it spots potential prey, at which time it springs into the air in pursuit. Nunbirds typically join large mixed foraging flocks and often follow army ant swarms.

Puffbirds are large-headed, heavy-bodied birds so named for the puffed appearance of their feathers. Though some species are boldly patterned in black and white, most species are brownish or tan. Their cryptic plumage in the shaded forest understory makes them easy to overlook as they perch motionless on a branch. The white-whiskered or brown puffbird (*Malacoptila panamensis*) is a common bird of the forest understory from southern Mexico through Ecuador. It is dark brown above and has a tan breast with brown streaking. Close examination reveals red eyes and white feathering around the bill. Higher in the canopy is the white-necked puffbird (*Notharctus macrorhynchos*), a larger bird with bold black and white plumage that ranges all the way from southern Mexico to Argentina. Both species have

large rictal bristles, hairlike feathers around the base of
the bill, which probably aid in capturing aerial insects.
These two puffbird species are generally segregated vertically, the white-whiskered in the understory and the
white-necked in the canopy. Such a distribution may reflect the outcome of both specialization for food capture
(canopy insects are not the same as those of the understory) as well as interspecific competition (since each species inhabits a different vertical area, they do not directly
compete with each other).

Nunbirds and puffbirds excavate nests in termite
mounds or in the ground. A puffbird pair seems undisturbed by the presence of an observer when the two birds
excavate a termitary and tolerate termites crawling over
them as they incubate. Those that burrow make very long
nest tunnels (Skutch 1983).

BARK DRILLERS AND PROBERS—WOODPECKERS AND WOODCREEPERS

Woodcreepers (family Dendrocolaptidae, order Passeriformes) are bark probers, and woodpeckers (family Pici-
dae, order Piciformes) both probe and drill bark. The
world's woodpeckers are treated by Short (1982), and
Skutch (1985) provides a general natural history of woodpeckers.

Woodpeckers occur globally (except Australia) wherever there are trees, and, thus, many are temperate zone
species. Neotropical woodpeckers vary in size from the
14-inch ivory-billed types (genus *Campephilus*) to the diminutive 3.5-inch piculets (genus *Picumnus*). The world's
largest woodpecker is the 22-inch imperial woodpecker
(*Campephilus imperialis*), now highly endangered, but
which formerly ranged through montane pine forests in
Mexico. Tropical woodpeckers range in color from bold
black with red crest, to greenish olive, to soft browns and
chestnut. Some have horizontal black and white stripes on
their backs with varying amounts of red on the head. One
species, the brilliant cream-colored woodpecker (*Celeus*

flavus) of northern Amazonia, is bright yellow-buff with brown wings and a black tail.

Woodpeckers hammer into bark and extract insects, mostly larval, by using their extremely long, extrusible, barbed tongues. They hitch up tree trunks, their bodies supported by stiff tail feathers that act as a prop.

Neotropical woodpeckers excavate roosting and nesting cavities that are often usurped by other species. Skutch (1985) observed a group of collared aracaris (*Pteroglossus torquatus*) easily evict a pair pale-billed woodpeckers (*Campephilus guatemalensis*) from their nest cavity. Skutch also reports that two tityra species steal cavities from several woodpecker species. Skutch portrays the 7-inch tawny-winged woodcreeper (*Dendrocincla anabatina*), which attacks and forces several woodpecker species from their cavities, as "the most consistently aggressive bird that I have watched in tropical America."

Woodcreepers look superficially like woodpeckers but bear no close relationship to them. The similarity is a case of evolutionary convergence (see toucans, above) brought about by similar ecologies. Like woodpeckers, woodcreepers have stiff tail feathers that prop them vertically against a tree trunk. They tend to climb upward, spiraling around the trunk. Woodcreepers evolved from the ovenbirds (Furnariids, see following section) and all have become bark-probing specialists. Woodcreepers feed quite differently from woodpeckers. A woodcreeper typically lands at the base of a tree and methodically spirals upward around the trunk, probing into crevices, poking its bill into epiphytes, and generally removing insects, spiders, and even an occasional tree frog. They rarely peck into the trunk, instead using their long bills as forceps to pick off prey. Woodcreepers may also join mixed flocks that follow army ant swarms (see below).

Like their evolutionary cousins, the ovenbirds, woodcreepers are colored soft shades of brown. Many have various amounts of yellowish white streaking on breast, head, and back. The overall size of the bird, its bill size and shape, and its steaking pattern, usually separates one

Three woodcreepers, showing differences in bill and body size.
Wedge-billed (upper right), red-billed scythebill (center left),
and strong-billed (lower).

species from another. The smallest is the 6-inch wedge-billed woodcreeper (*Glyphorynchus spirurus*), which has a very short but sharply pointed bill. The largest, the strong-billed woodcreeper (*Xiphocolaptes promeropirhynchus*), reaches just under 13 inches in length and has a long thick and straight bill. Among the oddest of the group are the scythebills (genus *Campylorhamphus*), whose extremely long, downward curving bills are used to probe deeply into bromeliads and other epiphytes.

Many woodcreepers are ant followers, joining antbirds and motmots to feed on insects and other animals disturbed by the oncoming army ants. With differing body sizes and bill shapes, several species of woodcreepers coexist and feed with little or no apparent competition. At one army ant swarm in Belize, I observed six wood-creeper species. Two were large, three medium-sized, and one was small.

Woodcreepers are common not only in rain forests but also along forest edges and disturbed jungle. Some species are highly vocal, their songs consisting of pleasant melodious whistled trills.

FOLIAGE GLEANERS—OVENBIRDS

The ovenbirds (family Furnariidae, order Passeriformes) are "little brown birds" of the American tropics. All ovenbirds are generally nondescript, their plumage basically brown, tan, buffy, or grayish. Identification of individual species can be very difficult, since differences among species are often subtle and hard to see in the field. This highly diverse family occurs not only in low-land forests but in all types of habitat ranging through Patagonian pampas, Andean paramos and puna, and coastal deserts and seacoast. The family takes its common name, *ovenbird* from several species that construct oven-like, dome-shaped mud nests. Not all ovenbirds build such structures. Some nest in natural cavities or in mud banks, and some make basketlike structures of twigs and grass. The evolutionary trends of furnarids are analyzed by Fedducia (1973).

Ovenbird species have among the oddest common names of any birds. One may encounter a xenops, a recurvebill, a foliage-gleaner, and a leafscraper (not to be confused with a leaftosser!). There are also woodhaunters, treehunters, treerunners, palmcreepers, and earthcreepers (not to be confused with streamcreepers!). There are barbtails, spinetails, tit-spinetails, softails, and one thistletail. Finally, there are thornbirds, miners, cinclodes, horneros, and canasteros. Good luck sorting out ovenbirds.

All ovenbird species are basically insectivorous, but as a family they do not show the bill diversification that is so evident in woodcreepers. Rather, ovenbirds tend to be habitat and range specific and develop specialized feeding behaviors. Some, like the ground-feeding leafscrapers and leaftossers methodically probe among the litter. Their bodies are chunky and almost thrushlike in shape. Others, like the slender foliage gleaners search actively

Spinetail, a member of the large family of tropical ovenbirds, the furnarids.

among the leaves, ranging throughout canopy and understory. Spinetails dart quickly from bush to bush while the small xenops (page 51) hangs chickadeelike as it searches the underside of a leaf.

ANT FOLLOWERS—THE ANTBIRDS

Antbirds (family Formicariidae, order Passeriformes) include the antbirds, antshrikes, antwrens, antvireos, antthrushes, and antpittas. They take their family name from the army ant following behavior of some species. However, not all antbirds follow army ant swarms. Some never do, some occasionally do, and some virtually always do. This latter group is often termed the "professional antbirds." Like the ovenbirds, antbirds have been given odd-sounding common names.

Antbirds are much more colorful than ovenbirds, with many sexually dimorphic species. Males are often boldly patterned in black and white. Some, like the very common barred antshrike (*Thamnophilus doliatus*) (page 51), are zebra-striped. Others are grayish black with varying amounts of white patterning on wings, breast, and flanks. Still others are chestnut or brown. Females tend to be rich brown, tan, or chestnut. Some antbirds have an area of bare blue or red skin around the eye and in some species iris color is bright red.

Most antbirds are foliage gleaners, picking and snatching insects, from foliage, and some snatch insects on the wing. They forage at all levels from the canopy to the litter on the forest floor. They typically form mixed species flocks with other birds and various antbird species tend to feed at specific heights above the forest floor. Mixed flocks of up to fifty bird species move through Brazilian lowland forests, of which twenty to thirty species may be antbirds. Certain species such as the flycatching antshrikes (genus *Thamnomanes*) occupy the role of "central" species in the flock (Willis and Oniki 1978). These antshrikes are highly vocal and act as sentinels, warning the others of impending danger should they spot a forest falcon or other potential predator (see above, tangers). Willis and Oniki describe the relationship among the various mixed flock species as a "casual mutualism." They each benefit somewhat from the each other's presence.

There are twenty-eight species of professional ant-following birds, each of which makes its livelihood by capturing arthropods scattered by advancing fronts of army ants. In addition, other species frequently, but not always, can be found accompanying the ants. There are even some butterflies that associate with army ants to feed on the bird droppings (Ray and Andrews 1980)! In northern Amazonia, the white-plumed antbird (*Pithys albifrons*) is among the commonest professional antbirds. This bird is unmistakable, its face dominated by a tall white crest, its head black, its back and wings blue-gray, and its breast and tail rich chestnut in color. The spotted antbird (*Hylophylax naevioides*), the bicolored antbird (*Gymnopithys leucaspis*), and the black-faced antthrush (*Formicarius analis*) are among the most devoted ant followers in Central America. Where these three are found together, there are surely army ants about (Willis 1966, 1967).

Ant followers rarely feed directly on army ants. It is suspected that the high formic acid content of these insects deters birds from eating them. Instead, antbirds feed on anything from insects to small lizards scared up by the oncoming ant columns. Two army ant species, *Eciton burchelli* and *Labidus praedator*, are the ants most frequently followed. Birds such as woodcreepers, ovenbirds, motmots, certain tanagers, and other "less professional" antbirds come and go as part of the ant-following avian assemblage, but the professional antbirds always stay with the ants. Only when breeding do they become territorial and cease to follow ants for a time. Even then, they will quickly orient to army ant swarms within their territories.

Species such as the spotted and bicolored antbirds feed actively in trees and undergrowth, while the black-faced antthrush walks sedately on the forest floor. With the stature of a small rail, the black-faced antthrush walks with its short tail cocked upward and head held up and alert. It is easy to imitate its whistled downscaled "chew, chew,

chew, chew" call. In Trinidad, I called one almost to my feet as I whistled and it answered.

Antbirds tend to mate for life. Both male and female are active nest builders (Skutch 1969). One species, the ocellated antbird (*Phaenostictus mcleannani*), forms clans. Sons and grandsons of a pair return to the breeding territory with mates to form clans and a clan will occasionally attack another intruding clan (Willis and Oniki 1978). Antbirds also sometimes intimidate migrant thrushes that attempt to gather at antswarms (Willis 1966).

Along Rivers and Streams

Riverine habitats include flowing waters as well as bordering swamps and marshes. As rivers cut through rain forests they provide an excellent vantage point for observing bird activity. From a boat lazily moving downriver you can watch three and occasionally four species of vultures soaring overhead, observe hawks, caracaras, and falcons perched in riverside trees, and see parrots ranging from frantic flocks of small canary-winged parakeets (*Brotogeris versicolorus*) to the larger and more sedate macaws. Groups of social flycatchers are conspicuous, and various species of swifts and swallows skim above the water pursuing insects. Many remarkable Neotropical birds tenant streamsides and riverine habitats.

HOATZIN

Perhaps most unique among riverine species is the hoatzin (*Opisthocomus hoatzin*). This extraordinary bird is found only in South America along slow meandering streams that comprise part of the Amazon and Orinoco basins. Hoatzins roughly resemble chickens in size and shape. However, their overall appearance suggests a primitive, almost prehistoric, bird. A hoatzin is somewhat gangly, its body chunky, its neck slender, its head small. The face is not feathered but rather consists of bright blue bare skin surrounding brilliant red eyes. A ragged crest of feathers adorns the bird's head. Adding to its antediluvian appearance is the bird's subdued plumage of soft browns with rich buff on breast and wings. Hoatzins are weak fliers, a feature that contributes to their primitive appearance. No one who sees a hoatzin ever forgets it.

Though originally believed to be related taxonomically to chickens, hoatzins are now considered most closely related to cuckoos, order Cuculiformes (Hilty and Brown 1986). The species is the only member of its family, Opisthocomidae. A hoatzin's unique appearance is complemented by an unusual diet, unusual breeding system, and unusual juvenile behavior. Hoatzins feed exclusively on leaves from plants of the arum family (i.e., philodendrons), which they swallow and grind into a large bolus in their oversized crops. The bolus slowly ferments and is

Adult hoatzin.

digested. The odd amalgamation of partially decomposed leaves gives the bird an unpleasant musky odor, a beneficial characteristic since it renders the flesh distasteful to native hunters. However, monkeys have no apparent aversion to the odor or (presumably) the taste of hoatzin and routinely prey on it.

Hoatzins are communal breeders, and anywhere from two to seven birds cooperate in a single nesting. The pair responsible for the eggs is usually assisted by nonbreeding birds called "helpers." Studies by Stuart D. Strahl (1985) have shown that nests with helpers are considerably more successful at fledging young than nests lacking helpers. The helpers aid in incubation and feeding young, enabling the juvenile birds to grow more quickly and thus reduce their vulnerability to predators. Their streamside nests are quite crude, consisting of a cluster of thin sticks so loosely constructed that the eggs are usually visible from beneath.

Baby hoatzins bear a superficial resemblance to *Archaeopteryx*, the first bird, whose fossilized remains established that birds evolved approximately 100 million years ago during the Mesozoic era, when dinosaurs flourished. Young hoatzins possess claws on their first and second digits, enabling them to climb about in riverside vegetation. Juvenile hoatzins swim and dive efficiently. Should they be faced with danger, they escape by dropping from the vegetation into the water. When danger passes they use their wing-claws to help in climbing back on vegetation. Wing-claws were also present on *Archaeopteryx*, though no one suggests that the resemblance between the modern hoatzin and the first bird is other than coincidental. Young hoatzins lose their wing-claws as they attain adulthood.

SUNBITTERN

Stalking along quiet riverbanks, the heron-shaped sunbittern (*Eurypyga helias*) hunts fish, amphibians, crustaceans, and insects, which it captures by striking quickly, using its long neck and spearlike bill. With a sharp white

Juvenile hoatzin, climbing, using claws on wings.

Sunbittern.

line above and below the eye, and complexly patterned plumage, the sunbittern resembles the sunflecked forest interior. When displaying, it spreads its wings, revealing bright chestnut, yellow, black, and white linings that give the bird its name. Its legs are bright red. Like the hoatzin, the sunbittern is the only species in its family, Eurypigidae (order Gruiformes).

SCREAMERS

Three species of screamers (family Anhimidae, order Anseriformes) are found along slow rivers, swamps, and marshes throughout South America. Screamers are most closely related to ducks and geese. I observed the horned screamer (*Anhima cornuta*) in the Peruvian Amazon. Nearly the size of a turkey (which it vaguely resembles), it is a shiny black bird with thick legs, large unwebbed feet, and a smallish chickenlike head. The horned screamer is named for both its long feather quill that tops off its head and for its loud piercing call. Screamers, though bulky birds, are excellent fliers and frequently perch in riverside trees. They are unique in their possession of a layer of air between their skin and muscle, and the buoyancy provided by this "inner tube" may aid them in soaring.

Horned screamer.

Look closely at the vultures overhead. There may be a screamer or two soaring among them.

STORKS, HERONS, AND EGRETS

Other tropical water birds are prone to soar high above rivers, marshes, and wet savannas. Anywhere from the southern Amazon through Central America, storks (family Ciconiidae, order Ciconiiformes), such as the wood stork (*Mycteria americana*), maguari stork (*Euxenura maguari*), and jabirou (*Jabirou mycteria*), can be seen making lazy circles overhead during the heat of the tropical day. The jabirou is among the largest storks, topping 4 feet in height. It is all white on wings, tail, and body with a bare-skinned black neck at the base of which is a bright red patch. Its bill is long, thick, and slightly upturned.

Herons and egrets (family Ardeidae, order Ciconiiformes) are also common in marshland and riverside throughout the tropics. The long-necked herons and white-plumaged egrets are more slender than storks and fly with their necks held in an S-shaped curve. Storks fly with outstretched neck and head.

BOAT-BILLED HERON

The boat-billed heron (*Cochlearius cochlearius*) is an odd inhabitant of mangrove swamps and riverbanks, named for its extraordinarily wide, flattened bill. Taxonomically unique, it is the only member of the family Cochleariidae (order Ciconiiformes). Colonies of boat-billed herons leave their roosts at night and feed individually along rivers and marshes. The function of their seemingly oversized bill remains largely unknown, though the bill may be touch-sensitive, aiding the bird in searching for creatures inhabiting mud. They also feed on frogs, fish, and crustaceans. The species was believed to be most closely related to the night herons, and, indeed, the boat-billed heron somewhat resembles the widespread black-crowned night heron (*Nycticorax nycticorax*) in overall plumage as well as in nocturnal habits. However, Charles Sibley and Jon Ahlquist (1983), examining the DNA of the boat-billed heron, have found it to be no more closely related to night herons than to day herons. Its early evolutionary history remains elusive.

JACANAS

Eight species of jacana (family Jacanidae, order Charadriiformes) use their elongate, unwebbed toes to delicately walk atop lily pads searching for arthropod food throughout the world's tropical marshlands and riversides. Two species, the northern jacana (*Jacana spinosa*) and wattled jacana (*J. jacana*), are Neotropical. Both are chicken-sized, blackish birds with dark rufous wings that reveal bright yellow patches when the birds fly. The northern jacana is one of the few birds of which only males incubate the eggs and any female will mate with several males.

KINGFISHERS

The 15-inch ringed kingfisher (*Ceryle torquata*) is abundant and conspicuous along the Amazon as well as other Neotropical rivers and coastal areas. All kingfishers (family Alcidinidae, order Coraciiformes) have large heads and bills and make their livings plunging headfirst into streams and rivers in pursuit of fish. The ringed is by far the largest of the five Neotropical kingfisher species. Bluish gray above with a rufous breast and white neck, its loud rattling call, large size, and distinct color pattern make it easy to recognize. Like all kingfishers, the ringed is a swift and direct flier.

The other four kingfisher species are quite similar to each other in appearance but differ in size. From largest to smallest, there is the 11-inch Amazon kingfisher (*Chloroceryle amazona*), the 9-inch green-and-rufous kingfisher (*C. inda*), the 8-inch green kingfisher (*C. americana*), and the 5.5-inch pygmy kingfisher (*C. aenea*). All of these, plus the ringed, may be spotted zooming along the same stretch of river (though the green-and-rufous fails to get into northern Central America). Each of the four is iridescent green above with various amounts of rufous on the upper and lower breast.

The kingfisher assemblage is yet another of the many examples of size gradation among related tropical bird species. As was discussed with flycatchers (page 141), the differing sizes may permit each species to specialize, taking a certain size range of fish. The degree of interspecific competition is, therefore, kept rather minimal, and the various species coexist within the same ecological community. Since captured fish are swallowed whole, it is hard to imagine that the diminutive pygmy kingfisher would ever be able to consume a fish that would comprise a square meal for a ringed kingfisher. Indeed, I have watched a pygmy kingfisher plunging into a shallow rainwater pool after tiny tadpoles and insects. Ringed kingfishers routinely eat fish larger in body size than a pygmy kingfisher!

Kingfishers tunnel nests in soft riverbanks.

JACAMARS

The sixteen jacamar species (family Gabulidae, order Piciformes) are exclusively Neotropical. Like kingfishers, they excavate nest tunnels in banks, hence their frequency along rivers. They are also commonly encountered both in forests and along forest clearings. Some of forest-dwelling species excavate nest holes in termite colonies. Jacamars are slender birds with rather long tails and long, sharp bills. They sit motionless on exposed limbs and sally forth to snap insects out of the air. Two basic color forms occur: one is metallic green above with rufous on the breast (a pattern also encountered among the kingfishers), and the other is dull blackish brown with white on the lower breast and belly.

Birds of Prey

Birds of prey are diverse and abundant in the Neotropics. They range in size from the tiny bat falcon and pearl kite to the majestic harpy eagle. Open areas, such as savannas, are excellent for searching out many of the larger species, since they soar on thermal currents rising from the hot ground. Inside forests, birds of prey can be elusive. Many, such as the forest falcons, sit motionless on a branch waiting for an opportunity to attack would-be prey.

KITES

Eleven species of kites gracefully skim Neotropical skies searching out small animals such as mice, birds, lizards, and arthropods. Kites have sharply hooked bills, a trait particularly evident in the snail kite (*Rostrhamus sociabilis*) and the hook-billed kite (*Chondrohierax uncinatus*). The snail kite specializes on one food source, the large marsh snail *Pomacea*, which it adeptly removes from the shell with its sharply hooked bill. Another common kite is the black-shouldered (*Elanus leucurus*), often seen hovering over open fields and savannas seeking its small animal prey. The most graceful flier among the kites is the swallow-tailed (*Elanoides forficatus*), a slender black and white kite with a deeply forked tail. The 9-inch pearl kite *Gampsonyx swainsonii* is one of the smallest tropical birds of prey. Mostly black with white underparts, it has a buffy forehead and face, and a white or rufous neck. Like most kites it frequents savannas.

HAWKS, FALCONS, AND CARACARAS

Forty species of hawks (family Accipitridae, order Falconiformes) can be found in the Neotropics. Included here are a mere sample.

The crane-hawk (*Geranospiza caerulescens*) is a slender blackish gray inhabitant of wet savannas, mangroves, and swamps. The bird has exceptionally long, bright orange legs. It feeds both on the ground and in trees and has been reported to assume odd postures, such as hanging upside down, when probing epiphytes and branches for amphibians and reptiles. Another savanna species is the savanna hawk (*Heterospizas meridionalis*), which tends to be seen walking about on the open ground on its long yellow legs. It is largely rufous, with black tail and wing tips and dark barring across its breast.

The white hawk (*Leucopternis albicollis*) is apt to be seen soaring on warm thermals over forests. As its name implies, it is virtually all white but for a black band across the tail and black on the wings and around the eyes. Other soaring hawks include the common black hawk (*Buteogallus anthracinus*) and great black hawk (*B. urubitinga*). Both of these birds are almost all black but for white tail bands.

The black-collared hawk (*Busarellus nigricollis*) is one of the most elegantly plumaged Neotropical birds of prey. Warm rufous in overall color, it has a white head and black throat, black outer wings and tail. This hawk has a distinct shape when seen overhead, because its wings are quite wide and tail short. Found around marshes, it feeds mostly on fish.

The roadside hawk (*Buteo magnirostris*) is very well named. The hulking shape of this grayish rufous hawk can be seen perched on cecropias, palm trees, ceibas, and telephone poles all along tropical roads. This very abundant species is also highly variable in plumage, and thirteen races have been recognized.

Falcons are small, speedy birds of prey known for their aerial agility. With long tail and sharply pointed wings, falcons are quick to pursue and capture rodents, small birds, and insects. One species, the diminutive bat falcon (*Falco rufigularis*), specializes in capturing bats at dawn and dusk. It is largely dark blue, with a white throat and orange on the thighs and lower belly.

The laughing falcon (*Herpetotheres cachinnans*) is often seen perched atop a snag along a forest edge, cleared field, or savanna. Very buffy on the head, neck and breast, the laughing falcon has dark brown wings, back, and tail, with a black band through the eyes and around the back of the neck. Named for its penetrating loud call, these birds prey on snakes and other animals spotted by patiently sitting for long periods.

Forest falcons (genus *Micrastur*) are grayish falcons that

skulk inside forests and are generally difficult to find. They are very inconspicuous, sitting motionless in the deep forest shade.

The yellow-headed caracara (*Milvago chimachima*) is common and conspicuous, often seen in groups along rivers and forest edges. Caracaras, like vultures, feed on carrion and, hence, are frequently encountered along roadsides. Yellow-headed Caracaras are slender birds, buffy yellow on head, breast, and belly with blackish brown wings and tail.

HAWK-EAGLES AND EAGLES

The largest Neotropical birds of prey are eagles and hawk-eagles. There are three species of hawk-eagles, each of which has a crest atop its head. The ornate hawk-eagle (*Spizaetus ornatus*) has a bright orange neck and a tall black crest. The black hawk-eagle (*S. tyrannus*) is uniformly dark, and the black-and-white hawk-eagle (*S. melanoleucus*) is black above and white below. Hawk-eagles are soaring hawks, usually seen above the canopy making circles high overhead.

The rare harpy eagle (*Harpia harpyja*) is among the most magnificent of Neotropical birds. This huge predator is 3.5 feet tall, with thick, powerful legs. Mostly gray on face and belly, the wings, back, and upper breast are black. The head sports a tall blackish gray crest. No bird of prey approaches the size of a harpy. Nonetheless, it is secretive, tending not to soar, and, thus, is difficult to see well. Strictly a forest dweller, now unfortunately rare over most of its range, the magnificent harpy eagle is a top prize for anyone seeking unique tropical birds. Harpys feed on monkeys and sloths.

OWLS

Owls (order Strigiformes, families Tytonidae [barn owls] and Strigidae [typical owls]) are nocturnal birds of prey. Over two dozen species occur in the Neotropics. Below are several of the most wide-ranging.

The spectacled owl (*Pulsatrix perspicillata*) is the largest

Adult harpy eagle.

Neotropical owl, reaching 19 inches in length. It is buffy yellow on the lower breast and belly, with dark brown back, wings, tail, and head. A dark brown band crosses its upper breast, and its bright yellow eyes are highlighted by white, giving the bird its name. Spectacled owls make a deep hooting sound and can be easily attracted to tapes of their voice.

The black-and-white owl (*Ciccaba nigrolineata*) is just that. Its breast, belly, back, and face are barred black and white. This species is also responsive to tapes of its voice.

The mottled owl (*Ciccaba virgata*) is warm brown and tan with dark brown eyes. Unlike its black and white relative, it does not tend to be attracted to tapes.

During the daytime, it is not uncommon to see a small pygmy-owl (genus *Glaucidium*) with its bright eyes staring as it perches atop a snag. Several species of these 6-inch owls occur in the tropics, but the most common is the ferruginous pygmy-owl (*G. barasilianum*), so named for its reddish brown plumage.

North American Migrant Birds

During autumn, 332 species, or 51% of all North American migrant bird species, fly to wintering areas in the Neotropics (Rappole et al. 1983). The majority winter in Central America, but many also winter in South America and the West Indian Islands. The abundance of North American migrants is high in the tropics from November through March. Not only are there many yearling birds in addition to adults, but the actual land area of Central America is less by about a factor of eight, compared with available nesting area in North America. Migrants ranging from Swainson's Hawks (*Buteo swainsoni*) to Least Flycatchers (*Empidonax minimus*) are packed into tropical America, and their abundance is evident.

Many North American migrants are essentially tropical in origin. Tyrannid flycatchers, hummingbirds, tanagers, orioles, and wood warblers all originally evolved in the Neotropics, and migrant species represent the relatively few that ventured northward, extending their ranges, perhaps because the northern summer presents an abundance of proteinaceous insect resources for the rearing of young, plus the availability of nesting sites.

Brushy successional areas are habitat for many species such as gray catbird (*Dumetella carolinensis*), northern yellowthroat (*Geothlypis trichas*), and yellow-breasted chat (*Icteria virens*). Rain forests provide habitat for wood thrushes (*Hylocichla mustelina*), ovenbirds (*Seiurus aurocapillus*, not to be confused with the many species in the family Furnariidae—this ovenbird is a species of migrant wood warbler, family Parulidae), and American redstarts (*Setophaga ruticilla*), and other wood warblers. But many of these species also utilize successional areas. Many North American migrants eat a diet high in fruit while in the tropics. Northern and orchard orioles and scarlet and summer tanagers feed in cecropia and fig trees among mixed flocks of euphonias, Neotropic tanagers, and honeycreepers. Leck (1987) cites studies indicating that fruit availability is very high on second-growth habitats. Many researchers have noted that abundance of migrants is high in successional areas and young forests. Higher fruit availability, providing high calorie fuel for migrants, may be the reason why migrants favor such areas.

The degree to which North American migrants interact with Neotropical resident birds is by no means certain and probably varies considerably depending on the ecology of the birds. Edwin Willis (1966), working on Barro Colorado Island, noted that North American thrushes are prevented by resident antbirds from access to swarms of *Eciton*. Many other researchers, however, argue that North American migrants may compete relatively little, if at all, with resident species. Much research remains to be done on this question. Works by Rappole et al. (1983) and Keast and Morton (1980) each contain comprehensive accounts of recent research on Neotropical migrants.

My colleague William E. Davis, Jr. and I (Kricher and Davis 1986, 1987) have investigated winter site fidelity among migrant species in southern Belize. We have shown that such species as the wood thrush, ovenbird, Kentucky warbler (*Oporornis formosus*), and gray catbird occupy exactly the same locations from one winter to the next. Although these birds migrate north to nest, they return in the fall to precisely the same local wintering area used the previous year. We placed fine mesh nets (called mist nets) in selected locations in both successional areas and rain forest. We captured some of the same banded individuals in the same nets, at the same locations, over three succeeding winters. Wood thrushes for example, establish and defend winter territories, using vocalizations and body posturing. Each wood thrush has its own spot within the rain forest. Other researchers have also reported high site loyalties and winter territoriality for other migrant species (Rappole et al. 1983).

Though many North American migrant species inhabit disturbed areas and montane forests, many also inhabit mature rain forest, and the degree to which deforestation (see chapter 10) may be eliminating wintering sites is becoming of increasing concern. Some migrant species are now rare in North America, possibly due to loss of their wintering areas. The Kirtland's warbler (*Dendroica kirtlandii*) nests only in successional jack pine forests in Michigan. The bird was once probably more widely spread and could possibly occupy a larger nesting area today, but winter habitat loss in the Bahamas gives it no place where large populations could winter. The Bachman's warbler (*Vermivora bachmanii*) is an extremely rare nester in southern hardwood swamps. Loss of cane habitat in Cuba, where the bird winters, is thought to have been responsible for its populational demise. On the other hand, increase in second-growth habitat could actually favor such species as chestnut-sided warbler (*Dendroica pensylvanica*) and indigo bunting (*Passerina cyanea*). The effect of tropical deforestation on migrants eludes a simple answer. Like virtually all of tropical ecology, simple answers to complicted questions are simply not there.

Birds are a magnet that help draw visitors to the Neotropics. Some come to merely augment an already long life list of species, wanting to see more parrots, more tanagers, more hummingbirds, and hoping also for a chance to find the ever elusive harpy eagle. Others, following in the footsteps of Darwin, Wallace, and their kindred, investigate birds in the hopes of adding knowledge about the mysteries of ecology and evolution in this the richest of ecosystems. Opportunities abound for research topics. There are many areas of Neotropical ornithology that are poorly studied, hardly a surprise given the abundance of potential research subjects. Like all other areas of tropical research, however, bird study is negatively affected by increasingly high rates of habitat loss. This chapter was an attempt to convey the uniqueness and diversity of a Neotropical avifauna whose richness faces an uncertain future.

CHAPTER 7

A Rain Forest Bestiary

ALFRED Russel Wallace (1876) was deeply impressed by the animals he observed in the tropics:

> Animal life is, on the whole, far more abundant and more varied within the tropics than in any other part of the globe, and a great number of peculiar groups are found there which never extend into temperate regions. Endless eccentricities of form and extreme richness of colour are its most prominent features, and these are manifested in the highest degree in those equatorial lands where the vegetation acquires its greatest beauty and its fullest development.

In this chapter I will try to convey some of the wonder Wallace felt when he met the creatures that dwell within rain forests. I present here an array of Neotropical mammals, reptiles, amphibians, and invertebrates that the rain forest visitor is most likely to want to see.

MAMMALS

As a group, rain forest mammals tend to be secretive and nocturnal, making it a challenge to see them well. Unlike the game herds of the African plains, rain forest mammals do not stand out in the open for easy viewing but rather scurry through the canopy or over the forest floor well ahead of the naturalist. Many are mostly nocturnal. Still, by careful stalking or quiet sitting, it is often possible to obtain excellent views of mammals.

Monkeys

Neotropical monkeys all belong to a group named the platyrrhines, referring to the position of the nostrils, which tend to open at the sides in contrast with Old World monkeys, the catarrhines, whose nostrils are closely spaced and point downward. New World monkeys are less known for their nostrils, however, than for their tails. Many platyrrhines, such as the spider, woolly, and howler monkeys, have prehensile tails, which they use as a one-fingered fifth limb. In Ecuador, I watched a woolly monkey nonchalantly hang by its tail as it fed on fruits.

The sixty-four platyrrhine species have adaptively radiated to fill many ecological niches in the rain forest. There are large apelike monkeys (the spider, woolly, and howler monkeys), medium-sized "typical" monkeys (the capuchins), monkeys with bald faces (the uakaris), a nocturnal monkey (the douroucouli), small lemurlike monkeys (the marmosets), and small squirrellike monkeys (the squirrel monkey and the tamarins). New World monkeys are forest animals, avoiding savannas. There are no Neotropical equivalents of the baboons of the African plains. However, some monkeys prefer interior forest and some frequent the forest edge. Some are more common in successional or secondary forests than undisturbed forests. Taxonomically, although all are platyrrhines, the "typical" monkeys are placed in the family Cebidae and the tamarins and marmosets, sometimes called the "squirrellike" monkeys, are in the family Callitricidae. Marmosets are small, some of which, like the 6-inch, 3-ounce pygmy marmoset, are among the smallest primates. The largest Neotropical monkeys are cebids, and the largest of the group are the 20-pound howlers. Moynihan (1976) provides a general review of the natural history of Neotropical monkeys.

Capuchins (genus *Cebus*) range from Amazonia through southern Central America. There are four species, each 1 to 2 feet in length (excluding the 18-inch tail) and weighing from 2 to 9 pounds. All have prehensile tails. Troops move quickly through forests foraging for fruits, leaves, and occasional insects. One is apt to encounter a capuchin group anywhere from low tangle along the forest edge to the canopy. They have even been

Woolly monkey, showing use of prehensile tail.

reported from mangrove forests. Capuchins are basically cute monkeys, and, thus, they are popular choices for use by organ grinders and circuses. Research on Barro Colorado Island in Panama on *Cebus capucinus* has shown that capuchins are important seed dispersers of several tree species (Foster, cited in Oppenheimer (1982). Their diet varies seasonally, and they favor mature fruits when available (Oppenheimer 1982). They also eat young stems and

Capuchin.

leaves as well as some insects and other arthropods. Their habit of eating certain insect grubs such as bruchid beetles, which they remove from fruits, makes them beneficial to trees parasitized by these insects. Capuchin troops are noisy, and their sounds attract agoutis and collared peccaries, which feed on fruits dropped by the simians (Oppenheimer 1982).

Titi monkeys (genus *Callicebus*) resemble capuchins but

are a bit smaller. They are highly active, very skilled tree-top jumpers and feed on a wide variety of fruits, buds, and various arthropods (including spiders and millipedes). Titis seem to seek out thick jungle growth in which to feed and rest. They are confined to Amazonia.

Squirrel monkeys (genus *Saimiri*) are widely spread through Amazonia into Central America. There is probably only one species, *S. sciurea*, the common squirrel monkey, though taxonomists are not in total agreement about the group. A squirrel monkey does not resemble a squirrel. It is a bit smaller than a capuchin, but its tail is equally long, giving it the appearance of a little animal with an immensely long tail. The eyes are dark surrounded by white "spectacles" and a black nose and mouth. Ears are white. Body hair is grayish with rich rufous on the back, arms, and tail. Squirrel monkeys frequent gallery forests, lowland rain forest, and successional areas. They come around villages to feed on bananas, plantain, and citrus. They eat all manner of fruits as well as insects. Squirrel monkeys can be abundant. One team of researchers estimated between fifty and eighty per square mile (Moynihan 1976).

A most unusual beast is the night monkey or douroucouli (genus *Aotes*). This smallish 2-pound monkey is the only genuine nocturnal monkey on earth. It ranges from northern Argentina and Paraguay throughout the Amazon Basin and north into Panama. One look at its owllike, rounded head with immense dark eyes and its soft grayish brown pelage is enough to discern its nocturnal way of life. Families of night monkeys while away the daytime hours cuddled together in a hollow tree. At night they forage anywhere from almost ground level to the top of the canopy. They search for fruits, buds, insects, and, occasionally, nestling birds. Their loud calls, heard at night, keep the foraging troop together. They communicate by "body language"; positioning of the whole body rather than by facial expressions, a probable behavioral adaptation to life in the dark (Moynihan 1976).

As odd in appearance as the douroucouli is in habits

Night monkey.

are the uakaris (genus *Pithecia*). These medium-sized to
large monkeys, and their close relatives, the sakis, are
found only in the upper Amazon. Uakaris have long,
thick body hair, reddish in one species and silvery white
in another, and utterly bald faces and heads. In both species the bald head is bright red! They lack the prehensile
tail of most Neotropical monkeys. Like virtually all of
their kindred, they feed on fruits, but, unlike many species, they are quite uncommon. Consider yourself very
lucky if you spot one.

The sakis resemble uakaris in general body shape, but
they have hair on their heads, and their bushy tails are
much longer than uakaris (though also not prehensile).
Sakis are reported to be very skilled at leaping from tree
to tree, giving them the Spanish name *volador*, or "flier."
They also feed primarily on fruits and are found mostly
in well-developed rain forest. Sakis are not as uncommon
as uakaris.

Spider monkeys (genus *Ateles*) are common and widely
distributed from Mexico through the lower Amazon.

Four species occur, and they are generally geographically separated. All tend to be dark in coat color though light-colored subspecies occur. Spiders are often hunted by Indians and occasionally fall prey to the large harpy eagle (page 271). Spiders are rather large but quite slender (hence the name *spider*), weighing about 14 pounds. Their prehensile tail ranges from 20 to 35 inches in length. Troops of spiders typically consist of about eight adult males, fifteen adult females, and ten babies and juveniles. At any given time, four females will be either pregnant or in estrus. Bachelor male troops also occur. Often fewer animals are seen together, because troops frequently fractionate in a given area during the day, reassembling at their sleeping tree at night. Spiders forage together in the treetops, often quite actively. Their slender bodies adapt them well for graceful movement through the canopy, and they seem to prefer mature forest. Spiders move by brachiation, swinging arm over arm from branch to branch in a manner very similar to Old World gibbons. Studies on BCI have shown that 80% of their diet is fruits and 20% leaves (Moynihan 1976). Like many other monkeys they tend to strongly prefer young leaves (chapter 5, page 192).

Woolly monkeys (genus *Lagothrix*) are named for their thick woolly fur, which may be black, brown, grayish, or reddish. They are slightly larger than spiders, weighing up to 22 pounds, and in body shape they more closely resemble howlers (see below). Two species occur throughout the upper Amazon Basin, but neither reaches Central America. Woollies are relatively common and, thus, are frequently observed. They prefer tall forest but forage for fruits, upon which they feed almost exclusively, at varying heights, not being confined to the canopy. Like spiders, woolly monkeys have prehensile tails and are highly skilled arboreal acrobats.

Howler monkeys (genus *Alouatta*) are as large or larger than woolly monkeys and are scientifically the most well known Neotropical monkeys. Many studies have focused on howlers, including their various behaviors, their troop sizes, communications, territoriality, vocalizations, and feeding habits. There are six species, ranging in color from very blackish, to brownish, to quite reddish, and distributed from the Amazon Basin to Argentina and Paraguay in the south and Trinidad, Panama, and Costa Rica to the north. Howlers are named for their ferocious voice, which echoes through the rain forest sunrise and sunset like the roar of an enraged jaguar. Males have a throat sac acting as a resonator, dramatically amplifying their calls. Their infamous howling serves to mark the troop territory, and two troops will come to a mutual agreement about the real estate without ever meeting one another, because the howling carries for nearly a mile through rain forest. Males are 30% larger in body size than females. An average howler clan consists of three adult males, seven to eight females, and varying numbers of juveniles. Clans vary in size, however, ranging from four to thirty-five. Males are dominant over females, and young animals tend to be dominant over older animals.

Howlers specialize on a diet of leaves, with fruit and flowers making up only 30.7% of their diet. Kenneth Glander (1977) found that they consumed 19.4% mature leaves, 44.2% new leaves, 12.5% fruit, 18.2% flowers, and 5.7% leaf petioles (see chapter 5, page 192). This diet takes in substantial protein (mostly from mature leaves) and minimizes the amount of undigestable fiber and potentially toxic defense compounds ingested (Milton 1979, 1981). Howlers are more often heard than seen as they tend to remain well up in the canopy of mature rain forest. However, in many places they are common and relatively easy to find.

Katherine Milton (1982) found that howlers on Barro Colorado Island suffer a high mortality rate. Seasonal changes in food availability, plus periodic unpredictable shortages of high quality foods, stress the population and are probably major causes of mortality. Milton also found some monkeys heavily parasitized by botflies (page 57), though she noted that most monkeys seemed to survive their botfly wounds.

Marmosets and tamarins are diminutive monkeys, the gnomes of the rain forest, scurrying through the branches, peeking out from behind leaves often larger than they are. Both marmosets and tamarins are found in interior forest but seem to prefer forest edges and successional areas. They resemble hyperactive squirrels as they scatter about in the low tangles of branches. The name marmoset is taken from the French *marmouset*, meaning "small boy."

Marmosets (genera *Callithrix* and *Cebuella*) include the tiniest Neotropical monkey, the 6-inch pygmy marmoset of upper Amazonia. The first pygmy marmoset I saw was clinging to the bars inside a bird cage in a Lima garden. Unfortunately, they are frequently kept as pets. Pygmy marmosets feed on berries, buds, fruits such as bananas, and various arthropods. They also have the odd habit of "sap-sucking," which involves gnawing holes into a favorite tree trunk and drinking the oozing sap. Their lower incisors are unusually long, an aid in chewing holes to harvest sap (Goldizen 1988). All fourteen marmoset species have rather bushy tails and bushy ear tufts and cheeks, and most are brownish, buffy, or grayish in coat color, often with patches of black and white.

Tamarins (genera *Leontopithecus*, and *Saguinus*) are similar to marmosets. The golden tamarin is perhaps the most beautiful of the group, with glistening reddish gold fur thickly surrounding its black face. Like marmosets, they search mostly for insects and other arthropods but will not hesitate to raid banana plantations.

Tamarins and marmosets are unusual among primates for their flexible breeding systems. In some ways they seem to exhibit the reverse of the normal primate pattern. They tend strongly toward monogamy, though most primates are polygynous (Hrdy 1981). However, recent studies have shown that one female tamarin may mate with several males (polyandry) without creating aggression among the males (Abrahamson 1985). Tamarin groups normally consist of four to six adults, typically two or more females and several males. Females are aggressive toward one another, and one female does all of the breeding in the group. Several males mate with the alpha female, and males devote much energy, more so than females, to parental care.

The saddle-back tamarin (*Saguinus fusciollis*) displays both polyandry and cooperative breeding (Goldizen 1988). A dominant female will mate with several males. Juveniles are normally raised with the help of older, nonreproductive individuals (brothers and sisters) as well as reproductively active males. Why should some individuals help others raise young? One possible explanation, at least for this species, may be that the cost of parental care is so high that helpers are needed. Tamarins normally birth twins, and the combined weights of the infants can equal 20% of the mother's body weight. Infants must be carried as the troop moves on its foraging route. Males help carry infants, reducing the burden on the mother, who still must feed her twins with milk. But, why should helpers help? Tamarin troops tend to consist mostly of closely related individuals. Those individuals that help raise a close relative such as a brother or sister are promoting their own genetic fitness, since each sibling shares 50% of their genes. By helping raise twins, a brother tamarin is promoting one full set of his genes, a clear case of kin selection.

Rodents: Agouti, Paca, Capybara, and Others

Rodents (order Rodentia) are the most diverse order of mammals, and the Neotropics harbor some of the most interesting as well as the largest of the assemblage. Several are aquatic, others arboreal, and still others are burrowers.

Agoutis (*Dasyprocta punctata*) are one of the most common larger rodents. Diurnal, agoutis are apt to be encountered anywhere inside forests and are abundant from the Mexican tropics southward to northern Argentina. One of my students described an agouti (as well as its close relative the paca) very well by asking me, "What are those little piggy things running across the trail?" Agoutis do not look like "little piggy things" when seen closely, but, from a distance, their chunky 25- to 30-inch bodies, long legs, and delicate prancing gait give more of an impression of a small hoofed animal than a rodent. They are buffy brown with thin but rather long legs and a mouselike face. They are evolutionarily related not only to pacas (see below) but also to chinchillas and guinea pigs. One rodentlike characteristic of agoutis is that they eat by sitting upright on their hind legs, holding their food (usually a fruit or seed) with their front paws in a manner similar to mice and squirrels. Agoutis collect more seeds than they eat at once and bury the remainder in a widely scattered pattern. The wide pattern is possibly adaptive in protecting the agouti's cache from discovery by peccaries. During times of shortage, agoutis dig up their buried seeds (chapter 1, page 10). When chased by a predator, an agouti emits a high-pitched alarm bark.

When agoutis retire for the evening, pacas (*Agouti paca*) come out. Pacas are also common, sharing virtually the same range and habitat as agoutis, but they are less frequently seen because of their nocturnal habits. Days are spent resting in a burrow along a stream bank. A paca resembles an agouti in shape but has larger eyes and longitudinal white stripes and spots along the sides of its reddish-brown coat. Their bodies are larger and legs proportionally shorter than agoutis, and pacas weigh more, up to 22 pounds. Pacas fight among themselves with great vigor, the males biting at one another with their prominent rodent incisor teeth. When threatened by a predator, however, they retreat to water and remain immersed until out of danger (or air). Pacas, like agoutis, feed on fruit but also take leaves and other veggies.

The most piggy of the piggy things is the 4-foot-long, 120-pound capybara (*Hydrochoerus hydrochaeris*). Though they are usually found in groups of about twenty, herds up to fifty or a hundred of these aquatic rodents, the hippos of the Amazon, have been seen feeding on water lilies, leaves, bark, and sedges that line the Amazon and its

Capybara.

many tributaries, lakes, and swamps. Capybaras range throughout Amazonia and have recently been reintroduced into the Canal Zone of Panama. In many places, humans have hunted the "master of the grasses," (a translation of the word *capybara*), and populations have been reduced. A capybara is stocky, has a light tan coat, and thick legs. The hind toes are webbed, an obvious adaptation to the animal's aquatic habits. Their heads are squarish, and eyes, ears, and nostrils are located on the upper part of the head, a possible adaptation aid in being able to keep track of things while swimming. Their natural enemies are caymans and jaguars.

Another aquatic rodent, much smaller than the capybara, is the coypu (*Myocastor coypus*). This animal only tops the scales at 20 pounds, far less than its larger cousin, and it looks very much like a North American muskrat (*Ondatra zabethica*). A coypu has a long naked tail, which agoutis, pacas, and capybaras lack; they are all virtually tailless. Coypus burrow along lakes and rivers and are found only in the southern Amazon and Patagonia.

Spiny rats (family Echimyidae) are among the other Neotropical rodents of the rain forest. As the name implies, these animals have bristly hair and quills that can break off after inbedding in a predator that has come too close. Most spiny rats are burrowers, but some live in trees. All are highly nocturnal and hard to see. They are found throughout Amazonia but not Central America. Spiny rats are somewhat similar in appearance to porcupines.

There are also true porcupines in the tropics. The prehensile-tailed porcupine (*Coendou prehensilis*) ranges through rain forests of South and Central America. It is totally arboreal, climbing through the trees like either a slow monkey or a fast sloth. Covered with quills, it looks like a slender long-tailed version of its North American counterpart. Like many mammals, it is nocturnal, sleeping away the day in a hollow tree cavity. Porcupines in the tropics seem to feed entirely on vegetation, including leaves and fruits, though their diets are not well studied.

Other Neotropical rodents include squirrels (genus *Sciurus*), similar in appearance to, though darker than (and with a red tail), North American gray squirrels (*Sciurus carolinensis*), and a group of mice collectively called spiny pocket mice (genera *Heteromys* and *Liomys*). These little mice look very much like North American deer mice and white-footed mice (genus *Peromyscus*). They feed heavily on seeds.

Collared and White-lipped Peccaries

Peccaries are members of the huge even-toed ungulate order, the Artiodactyls. They closely resemble wild pigs but are in their own family, the Tayassuidae. They differ from pigs in that their upper canine teeth are extremely sharp and point straight downward, whereas in pigs these teeth curve outward as tusks. Peccaries also have a dorsal scent gland, located toward the posterior of their backs. This gland exits through a large and conspicuous opening that was once mistaken to be the animal's navel! The secretion of the gland is quite musky, and the animals rub their faces vigorously against one another's scent gland as a means of recognition and solidification of the herd. Peccaries are highly social animals and are rarely seen singly.

The most common and also the smallest of the two peccary species is the 50- to 65-pound collared peccary (*Tayassu tajacu*). This species is abundant in forests and savannas throughout tropical America as far south as Argentina and ranges northward through the Mexican deserts into the American Southwest. Collared peccaries form herds of anywhere from three to thirty or more. Their bristly hair is a mixture of black and gray as adults, but they are quite brownish as juveniles. The name *collared* is derived from a band of whitish hair that separates neck from shoulder. Their faces are very piglike, and their snouts have a hard fingernaillike rhinarium that acts as a trowel in rooting up vegetation. Herds of peccaries forage like pigs for roots, bulbs, and underground stems as well as leaves and fruits. They also eat arthropods and small vertebrates, if they can catch them. Loose soil from their rooting efforts is a common sight in rain forests along with the prints of their small cloven hooves. During dry season they often congregate at favored watering places. As they forage, they communicate with soft continuous grunts, but, should danger threaten, they emit a loud deep "Woof!" reminiscent of a large dog's bark. When cornered, they erect their bristles, chatter their teeth, and display their large canines. They put on quite an impressive show of threat but will not charge unless no escape is possible. Collared peccaries have an undeserved reputation for aggression. They are fundamentally peaceful, highly social animals (Sowls 1984). Should you encounter a band of them, give them a wide berth, and they will go about their business, leaving you totally intact.

The larger white-lipped peccary (*Tayassu pecari*) is more potentially dangerous than the collared. This animal, identified by white hair around its mouth, congregates in herds of up to a hundred or more and rarely is found outside rain forest. Its range is limited to the lowland forests of South America and southern Central America. Herds of white-lipped peccaries have been known to charge people who stumble upon them. Climbing a tree is the easiest escape route, though the peccaries may wait around beneath the tree for a while.

Richard A. Kiltie (1982) studied the feeding behavior of both peccary species in the Peruvian Amazon and calculated that white-lips can exert a bite force 1.3 times that

Collared peccary.

of collareds. In other words, white-lips can crack tougher fruits and seeds than collareds and, thus, can make use of food items unavailable to their smaller cousins. Kiltie speculated that the large herd size of white-lips may be related to diet. Some of the hardest fruits, the palms, drop many fruits at once, representing a temporarily abundant but highly patchy resource. In large herds, white-lips can find and exploit such a resource effectively. White-lips have a narrower rhinarium than collareds and do not dig as deeply for roots. They probably depend more heavily on hard fruits.

Both collared and white-lipped peccaries are extensively hunted by Amazonian Indians, comprising a major source of protein for the scattered settlements. Research conducted by William T. Vickers (1988) has challenged a long-held belief that the nomadic nature of Amazonian peoples is due to depleting local game supplies. In a ten-year study, he found that kill rates and hunting success remained high and that variations in hunt yield were not due to depletion by human hunters. Vickers attributed the nomadic nature of Amazonian settlements to other factors (page 95).

Tapirs

Tapirs (genus *Tapirus*) are Perissodactyls, or odd-toed ungulates, evolutionary relatives of rhinoceroses and horses. Only four species of tapirs occur in the world and three of them are in the American tropics (the other is in Indochina). Tapirs are stocky, almost hairless animals, brownish to black depending upon species, with a short elephantine proboscis and a dense but short mane of stiff hair on their upper neck. The mane probably aids the animal in making its way through dense undergrowth. Tapirs have an acute sense of smell and select food plants at least in part on the basis of odor. They eat only vegetable matter, including leaves and fruits of various species. Research on a captive tapir in Costa Rica indicated that most of the common rain forest plant species were unacceptable as food plants (Janzen 1983d). If true of free-living animals, tapirs could experience food shortages easily. Both tapirs and peccaries are widely hunted

Adult tapir and juvenile.

and, thus, tend to be wary. Tapirs also tend to be most active at night, and you will be very lucky to see one well.

Sloths, Anteaters, and Armadillos

Sloths and anteaters are among the most characteristic animals of tropical rain forests. Along with the armadillos, one species of which ranges into North America, they comprise the order Edentata, meaning "without teeth." Though anteaters do completely lack teeth, sloths and armadillos have peglike teeth on the sides of their mouths. Only front teeth are lacking.

The most commonly seen of the anteaters is the 2-foot-long tamandua (genus *Tamandua*). Two species occur, one in Central America and most of South America, and the other in the southern Amazon. Both look similar with long pointed snouts, formidable curved claws on the forelegs, prominent ears, and a long prehensile tail. The

Tamandua.

coat color is buffy brown with black on the back. Tamanduas are equally at home digging up ant nests on the ground or sampling the delicacies of termitaries in the trees. They excavate with their very sharp front claws and extract the insects using their extensable sticky tongues. Tamanduas don't eat just any ants. They tend to shy away from army ants and ponerine ants, both of which give nasty stings. When threatened, a tamandua may sit up on its hind legs and brandish its sharp curved claws. Tamanduas are largely solitary and are active day or night.

The giant anteater (*Myrmecophaga tridactyla*) is much larger than the tamandua and is totally ground-dwelling. Its body measures about 4 feet in length, and its huge bushy tail adds almost another yard. Its head is shaped like a long funnel with eyes and ears placed well back of the small mouth from which can protrude a 20-inch sticky tongue. The grayish black coat color is punctuated by a broad black side stripe lined with white, the entire animal terminating in an immensely thick ragged tail. Like the tamandua, the front claws are curved and sharp, an adaptation to digging into the very hardened ant and termite nests that contain dinner for the anteater. It ranges through Amazonia and southern Central America, though it is now rare throughout much of its range. The animal remains common in savanna areas in Venezuela. Also like the tamandua, the giant anteater will rear up and display its front claws if danger threatens.

A third anteater species is the silky anteater (*Cyclopes didactylus*). Smallest of the three, it measures only 1.5 feet in length. Don't count on finding one of these little creatures, for they are nocturnal and arboreal, climbing about in thick lianas. They have soft golden buffy fur, short snouts and claws, and a prehensile tail. Their large black eyes testify to their nocturnal habits. They seem to eat only ants.

The term *sloth*, when applied to a person, has come to mean sluggish, lethargic, dull, perhaps even dim-witted. Real sloths are probably all of these things. They do, however, have the tremendous advantage of staying around once you find one. If you care to watch a sloth move from one tree to another, a distance that would take a few seconds for a monkey, plan to spend about a day or so. Sloths are truly animals with slow motion lives.

The most common sloth is the three-toed sloth (genus *Bradypus*), the favorite of Charles Waterton (see page 114). Several species occur ranging throughout Neotropic forests. The three-toed sloth looks somewhat like a deformed monkey. It has shaggy tan-colored fur, long forearms and hind legs (but no tail), and a rounded face with very appealing eyes. It's sadly vacuous expression gives the impression that it entertains an actual thought only on alternate Tuesdays. The name derives from the three sharp, curved claws on each of its four feet, which serve as hooks as the animal hangs upside down from a branch, like an odd mammalian Christmas tree ornament. The easiest way to find a three-toed sloth is to scan the cecropia trees (chapter 2, page 53). Sloths sit motionless in cecropias often in the center-most part of the tree.

For years it was assumed that three-toed sloths eat only cecropia leaves, but a study by G. G. Montgomery and M. E. Sunquist (1975) on BCI showed differently. Sloths fed on leaves of ninety-six species of trees other than cecropia, though they did enjoy plenty of cecropia leaves. Interestingly, when sloths were in the crowns of these other tree species they were next to impossible to spot. Only by attaching tiny radios to the sloths were Montgomery and Sunquist able to locate them as they foraged in noncecropias. This is probably the reason why it was thought their diet was exclusively cecropias. They were just never seen in their other food trees! Further studies by Sunquist and Montgomery showed that sloths move to a different tree about every day and a half. They come to the ground only to urinate and defecate about once a week, which they do just at the base of a tree. They dig a small depression and cover their excrement, an operation requiring about thirty minutes. Montgomery and Sunquist suggested that this odd behavior may facilitate uptake of nutrients by the tree and, thus, enhance the growth of new leaves for the sloth. Another possible explanation is that because sloths defecate on the ground, predators cannot cue in on the odor of the sloth's fecal matter to find the sloth. Sloths have been estimated to have a population of from five to eight per hectare in Panama, a high population density. Because they have an extraordinarily low metabolic rate, they do not eat as much as their numbers and body size (2 feet long, up to 10 pounds) might suggest. They are relatively efficient digesters, having a complex and long digestive tube. Three-toed sloths have few predators, among them the rare harpy eagle.

The other kind of sloth is the two-toed sloth (genus

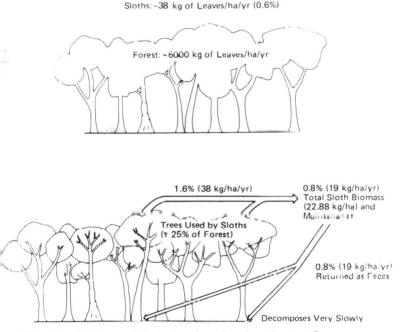

The role of three-toed sloths in mineral cycling. See text for details. From Montgomery and Sunquist (1975). Reproduced with permission of Springer-Verlag, Berlin, Heidelberg, New York, Tokyo.

Chloepus). The two-toed is similar in habits to its better known relative but is identified by its darker color and two, rather than three, prominent claws on the front feet and four, rather than three, on the hind feet. They are larger than three-toeds, weighing up to 20 pounds.

The largest of the sloths is now extinct. *Megatherium*, the giant ground sloth, inhabited the Americas during the Pleistocene, becoming extinct only within the past few thousand years, perhaps due to hunting pressure by humans migrating into their range (Krantz 1970; Long and Martin 1974). Generally resembling a gigantic version of the tree sloths, but approaching the size of a small elephant, *Megatherium* lived on the savannas and was protected from predators, at least as an adult, by its immensity. Charles Darwin discovered *Megatherium* fossils in South America and was deeply impressed by their similarily to living tree sloths. This observation was influential in later convincing Darwin to develop a belief in evolution.

The other members of the Edentata are the ubiquitous armadillos, of which there are many species. One (*Dasypus novemcinctus*) ranges into southeastern North America and is expanding its range northward. Armadillos are slow-moving ground dwellers whose hard bony skin protects them from most predators. When attacked, they curl up in a tight ball with their vulnerable soft parts tucked in. Mostly nocturnal, they are quite common especially in savannas, and they feed on a variety of insects and other arthropods.

Tropical Raccoons and Weasels

The familiar raccoon (*Procyon lotor*), raider of garbage cans throughout much of North America, ranges southward as far as Panama. Should you be driving along a Central American road at night or camping inside a rain forest, do not be shocked if this black-masked, ring-tailed beast makes an appearance. Raccoons are members of the family Procyonidae, which also includes the coatimundi, kinkajou, and olingo (see below). In addition to the common raccoon, the tropics boasts a species called the crab-eating raccoon (*P. cancrivorus*). This animal is quite similar in appearance to the common raccoon, but its neck fur grows forward rather than backward, and it lacks the thick underfur of the more northern species, though it is doubtful that you will get close enough to confirm this for yourself. Crab-eaters frequent mangrove swamps and other coastal areas and are generally nocturnal.

Coatimundis (*Nasua narica*) are familiar diurnal denizens of forests throughout the Neotropics. The species ranges northward into Arizona and New Mexico and southward to Argentina. Some populations occur in the high Andes, and others are found in deserts and savannas. Coatis usually travel in small bands mostly comprised of females and young. The males tend to be solitary except during breeding time. Coatis shuffle along, resembling streamlined raccoons, with highly pointed snouts, black and white face masks, slender, 25-inch-long grayish brown bodies, and a 13- to 28-inch slim tail, usually with faint rings (you must be close to see this) and usually held upright, like a cat holds its tail. Though their tail is not prehensile, they are adept at tree climbing and are as apt to be seen in trees as on the ground. They feed on all manner of things, including fruits, ground-dwelling invertebrates, and lizards and mice. I have seen them frequenting the garbage dumps and picnic areas at Tikal National Park in Guatemala.

The kinkajou (*Potos flavus*) is smaller than a coati and uniformly grayish tan, with an extremely long prehensile tail. Kinkajous are as nocturnal as coatis are diurnal. They scurry about the tree branches at night often making loud squeaking vocalizations, the banshees of the rain forest canopy. They can often be seen if you search with a flashlight with a strong beam that will penetrate the canopy at night. Kinkajous have forward-placed large eyes and wide, rounded ears and, thus, look a bit like monkeys. Omnivorous, they feed on fruits and small animals.

The Olingo (*Bassaricyon gabbii*), in face and body shape,

Kinkajou.

resembles a kinkajou, but it has a browner coat color and a ringed tail, which is not prehensile. Olingos are not well studied and are rarer than kinkajous, with which they share similar habits. They are nocturnal, rarely leave the trees, and feed on fruits and small animals.

The weasel family, Mustelidae, is represented in the tropics by several noteworthy animals. The most common is the 2-foot-long tayra (*Eira barbara*). Resembling a large mink or marten, the tayra is a sleek, blackish brown animal with a grayish face and an 18-inch, black bushy tail. It occurs throughout the Neotropics and, because it is diurnal as well as nocturnal, it's observed frequently. Tayras lack prehensile tails but are good tree climbers. They are quite omnivorous, feeding on rodents, nestling birds, lizards, as well as eggs, fruits, and honey. They are found in many different habitats as well as rain forests, including savannas and coastal areas. They raise young in a den usually located in a hollow tree trunk.

Tayra.

The two grison species (genus *Galictis*) resemble badgers. They are much smaller than the tayra and are slinky-looking animals with gray backs and dark undersides, with a white stripe behind the eye. Their tails are short and thick, colored gray. Like the tayra, they are not confined to forests but can be spotted in savannas and other open areas, and they are commonly encountered near human habitations where chickens and other tempting morsels are present. Strictly carnivorous, they feed on rodents and other small vertebrates. There are several species, and their combined ranges include Mexico through South America from lowland forest through the Andean foothills.

Among the aquatic mustelids, the largest is the giant otter (*Pteronura brasiliensis*). Just as the Amazon has its giant snake (anaconda), giant turtle (the arran), giant fish (arapaima), giant rodent (capybara), and giant water lily (Victoria), so has it a giant otter. This creature measures almost 5 feet in length, not counting its 2-foot long tail! With a sleek, reddish brown coat, the giant otter is distinguished by its four fully webbed feet and its tail, which is semiflattened, somewhat like a beaver's tail. Giant otters are social, and groups forage diurnally in the quiet waters of the Amazonian tributaries. Carnivorous, they feed on fish, mammals, birds, and other vertebrate prey.

If you think you see or smell a skunk in the tropics, chances are you are right. The spotted skunk (*Spilogale putorius*) occurs as far south as Costa Rica, the hooded skunk (*Mephitis macroura*) gets into Nicaragua, and the hog-nosed skunk (genus *Conepatus*), of which there are several species, ranges as far south as Patagonia and up into the Andes.

The Tropical Felines

Most people who travel to the American tropics for the thrills of seeing wildlife, when asked which animal they would most like to see, would probably name *El Tigre*, the jaguar. The lion may reign supreme on the African savanna, but in the Neotropical rain forest, *Felis onca* is the top cat. Though this 6-foot cat ranges from northern Mexico through Patagonia, it is now quite rare over much of its range. Only in the interior Amazon, remote montane forests, and other areas out of the immediate reach of hunter's gunfire does the 300-pound leopard of the New World remain unmolested. In central Belize, a reserve, the first of its kind, has been created specifically to preserve jaguars. Jaguars closely resemble leopards in spotting pattern but are generally heavier. Size varies considerably among individuals and some jaguars are known to top 400 pounds. The jaguar has no predators, with the rare exception of the anaconda, the sepentine giant of the Amazon (see below, page 310).

Jaguar are ecological generalists. These cats are found in lowland and montane forests, along rivers, in jungle, savannas, and coastal mangroves. They feed on many kinds of animals, including deer, tapirs, peccaries, sloths, capybara, giant otters, fish, birds, reptiles, even caiman. Largely solitary and basically nocturnal, El Tigre's footprints are seen far more often than the beast itself. A jaguar attacks its prey with a vigorous leap, quickly attempting to sever the neck vertebrae. The name *jaguar* derives from the Indian word *yaguar*, meaning "he who kills with one leap."

Much smaller and considerably more common than the jaguar are the ocelot (*Felis pardalis*) and its close relative the margay (*Felis wiedii*). The ocelot and margay are placed in the genus *Felis*, along with the common domestic house cat, which they resemble but for their fancier pelts. The big-eared, bright-eyed ocelot is about 3.5 feet in length with a thick 16-inch tail. A large individual weighs about 25 pounds. The margay is quite similar in appearance to the ocelot but slightly smaller in size. It is difficult to separate these species on the basis of a quick look because their spotting patterns are so similar. Both small cats are skilled tree climbers, though the margay seems the superior of the two. Margays are believed to be more nocturnal than ocelots and are, thus, less frequently

Ocelot.

sighted. Both animals are carnivores, feeding on anything from monkeys to insects. Like their domestic brethren, they spray to mark their territories. Jaguars, ocelots, and margays are the unfortunate victims of pelt seekers who make coats and then profits from killing these magnificent cats.

Of the various Neotropical cats, the jaguarundi (*Herpailurus yagouroundi*) is the most frequently seen. It is common and diurnal, often found in savannas as well as second-growth jungle. It is also the most frequently misidentified cat; many who are unfamiliar with it assume it to be a large weasel and not a cat at all. A jaguarundi looks superficially like a weasel because it is long and sleek, with proportionately short legs and a long tail. Its coat color varies from dark brown to gray to black, depending on the individual. It's a bit over 3 feet in length (not counting the 2-foot tail) and is the least arboreal of the cats, preferring to hunt from the ground. Like the other cats, jaguarundis are solitary.

The other large cat of the Neotropics is the mountain lion (also called puma, or cougar; *Puma concolor*), which is widespread in Central and South America as well as North America. It is an animal of the open savannas, forests, and mountains. It would be very much more common were it not for persecution by people. This large, light tan cat with a long tail is quite solitary and wide ranging.

The Marsupials

Most people associate marsupials, the kangaroos, wallabies, wombats, and bandicoots with Australia. Marsupials are mammals that give birth to very premature young that migrate to a pouch on the mother's abdomen where they attach to a teat and complete their development. Almost everyone has seen pictures of a mother kangaroo with her "joey" in her pouch.

Though Australia is the world's undisputed marsupial capital, until relatively recently, South America boasted a diverse marsupial community. Today, though there are many species of opossums living in the Neotropics, most of the original South American marsupial fauna is extinct. Around 3 million years ago a land bridge over the Isthmus of Panama provided a route for exchange and mixing of the North and South American faunal assemblages. A few species, like the durable opossum, migrated northward where it continues to expand its population today (Carroll 1988). However, many placental mammals migrated into South America, and some seem to have outcompeted or otherwise replaced the marsupials. For example, North American placental sabre-toothed tigers entered South America via the land bridge, and, relatively soon thereafter, marsupial sabre-toothed tigers became extinct (Carroll 1988). South America's marsupial fauna today is but a remnant of what it once was.

The familiar American opossum (*Didelphis marsupialis*) has hardly changed in appearance from its ancestors who roamed the planet 65 million years. The great carnivorous dinosaur *Tyrannosaurus rex* probably gazed on opossumlike creatures looking scarcely different from those that get flattened by autos on today's interstate highways. The fossil record indicates that the American opossum dates back about 35 million years, and its family dates to the Upper Cretaceous period, contemporary with the last of the dinosaurs (Carroll 1988). Superficially ratlike, with pointed snout and scaly naked tail, the American opossum weighs between 5 and 10 pounds and is largely gray with some black. The species ranges from eastern and parts of western North America throughout the Amazon Basin and into northern Patagonia. It inhabits almost any kind of terrestrial habitat other than desert and high mountains. Opossums are good tree climbers and often hang upside down, clinging by their prehensile tail. Totally omnivorous, the opossum will try eating almost anything. Its most unique behavior, "playing possum," is an act that feigns death when the animal is threatened.

In addition to the American opossum, the Neotropics hosts numerous other opossums. Most are nocturnal, but many are common and seen relatively often in daytime. There are furry little woolly opossums (*Caluromys*), tiny little mouse opossums (*Marmosa*), bushy-tailed opossums (*Glironia*), four-eyed opossums (*Metachirops* and *Metachirus*; which are four-eyed in name only!), short bare-tailed opossums (*Monodelphis*), little water opossums (*Lutreolina*), and the yapok, or big water opossum (*Chironectes*). Though these species are different sizes, they are all basically similar in anatomy, and the many species range throughout the various habitats of Central and South America. What this indicates is that opossums have undergone a successful adaptive radiation throughout the American tropics and are the survivors of what was for marsupials, a bygone era.

REPTILES AND AMPHIBIANS

The tropics are always associated with snakes. The fear of snake-infested trees and trails taints many peoples' views about the beauty of rain forest. How can you admire the scenery when you always have to be looking out

Mouse opossum.

for poisonous snakes? In reality, as I explained in chapter 2 (page 56), poisonous snakes are not frequently sighted. They tend to be secretive and nocturnal, and it's not easy to find them, even when you search diligently. There are, however, many species of snakes in the tropics, both poisonous and nonpoisonous, and as a group they are fascinating. Once people conquer their initial fears of serpents, I have seen them develop intense curiosity about them, followed by admiration of their beauty. This section will address that curiosity and will focus on other reptiles and amphibians as well.

Iguanas and Other Lizards

The prehistoric-looking common iguana (*Iguana iguana*) is a ubiquitous inhabitant of rain forests. These lizards are among the largest of their clan to occur in the Neotropics. They are quite greenish when small but become dark brownish black when full-sized. A large iguana can exceed 6 feet in length, but much of it is tail, which tapers into a slender whiplike tip. The face, with large mouth and wide, staring eyes, plus the antediluvianlike

Common iguana.

body of the reptile, provide a dinosaurlike countenance.
Two short spines adorn its nose above the nostrils, and a
loose membrane of skin called a dewlap hangs below its
throat. The head is flat and covered by heavy tuberclelike
scaling, and the neck and back are lined with short flexible spines. The legs sprawl alligatorlike to the sides, and
the feet have long toes with sharp claws. Iguanas do not
often hurry. They spend most of their time in trees, usually along a stream or river, into which they jump should
danger threaten. Excellent swimmers, they can remain
underwater for considerable time. When small they concentrate on insect food but feed more heavily on fruits
and leaves when full-sized. Should you encounter an
iguana, even a large one, you have nothing to fear. They
usually do not bite unless thoroughly harassed, they are
not effective scratchers, and they are nontoxic. The most
aggressive iguana I ever encountered was directing its
hostility at a rat. Both the mammal and the reptile were
contesting access to garbage dumped alongside the Amazon River in Iquitos, Peru. The iguana lost.

Iguanas are members of the large family Iguanidae,
which includes the many anolis lizards, basilisk lizards,
and ctenosaurs. Anolis lizards (genera *Anolis* and *Norops*)
are abundant. Some are bright green, some are brown,

and some are mixtures of both. Some can change color, rather like chameleons. During the heat of the day the sounds of these and other small lizards scurrying over dry leaves proceeds you as you walk along. Anolis have sharply pointed noses and large conspicuous dewlaps, which males use during courtship. They are skilled tree climbers and are often seen facing downward on a tree trunk with necks stretched out horizontally. They are also common on foliage. Small arthropods make up their diet.

The basilisk (*Basiliscus basiliscus*) also goes by the name Jesus Christ lizard because of its ability to run (not walk!) across water. With long toes on the hind feet, lined with skin flaps, these odd lizards run full tilt on their hind legs across small streams and puddles. They look even more like dinosaurs than iguanas do because they are adorned with elongate spiny fins on the back and tail. Catholic in terms of diet, they feed on invertebrates, vertebrates, and various fruits and flowers. Studies in Costa Rica indicate that basilisks are very abundant lowland animals. Estimates indicated that there are between two hundred and four hundred per acre (Van Devender 1983). Basilisks are found primarily in Central America.

Ctenosaurs (genus *Ctenosaura*), or black iguanas, are among the larger iguanid lizards. Ctenosaurs closely resemble iguanas but have a more banded pattern (though they can change pigmentation easily and pattern varies widely from animal to animal; some are dark, some light), and they tend to frequent drier areas such as open fields, farmyards, savannas, roadside edges, and coastal areas. They are adept burrowers and skilled tree climbers. Like iguanas, the larger ctenosaurs concentrate on vegetable food, though they are not averse to sampling such delicacies as bats, baby birds, and each other's eggs. Both ctenosaurs and iguanas are eaten by native peoples. Ctenosaurs are said to be the better tasting of the two. I hope this is true. I've had iguana.

Iguanas and ctenosaurs shift their diets from primarily arthropods to primarily vegetation, as they increase in body size. F. Harvey Pough (1973) hypothesized that this diet shift may be related to reptilian energetics and constraints of large size. Small lizards can be highly active and successful in catching small but scurrying prey such as beetles and spiders. Large lizards require more energy (simply because they are larger) but actually need less energy per gram of body weight (because large animals have slower metabolisms). They are probably not as well served by spending time and energy trying to capture fast-moving insects, none of which individually contain much energy. Plants don't move, require little energy to "capture," and, though harder to digest, more can be swallowed at a single sitting. Thus a vegetarian diet is more optimal for a large lizard. Even some of the medium-sized carnivorous lizards do not pursue prey, because such pursuit would expend more energy than would be contained in the prey item. These animals "sit and wait" and capture any slow moving arthropod (such as a caterpillar or grub) that happens to blunder past.

The lizard family Teiidae includes the largest of the South American lizards, the tegus (genus *Tupinambis*). Three species, the common tegu (*T. teguixin*), the northern tegu (*T. nigropunctatus*), and the red tegu (*T. rufescens*), each have large bodies, and they range from Central America through Argentina. The common tegu reaches a length of 55 inches, making it the largest lizard species of the Americas. Tegus live in forested areas, eating small animals. They are not averse to eating chickens and chicken eggs, and local farmers hunt them as a protein source. Some members of the Teiidae, the water teiids, are partly aquatic, with flattened tails that function well as paddles. The cayman lizard (*Dracaena guianensis*) of northeastern South America bears a resemblance to its namesake. This 4-foot lizard lives in *igapo* woodlands, floodplains that are flooded most of the year.

Geckos (Gekkonidae) are a major lizard family of the world's tropics. They are usually light with dark splotches. With suctionlike scales on their feet, geckos cling comfortably to smooth walls. They feed exclusively on arthropods and are nocturnal, inhabiting dwellings and living in harmony with humans. Geckos are considered valuable for their abilities to keep numbers of cockroaches and other vermin within tolerable limits. The name "gecko" comes from their loud calls, given only at night, often while hanging on a wall rather near where you're sleeping. It takes getting used to.

Snakes—The Constrictors

The world's largest snakes are constrictors (family Boidae). In the Old World, these snakes are called pythons, but in the Neotropics, they are the boas and anaconda. A few boas are found in Madagascar and the Indo-Pacific islands, but most are in the Americas. Boas are nonpoisonous, though their teeth are needle sharp, and their bite can be nasty. They capture and kill prey through constriction, a process whereby the serpent coils around its victim tightly enough to prevent it from breathing. Boas kill by suffocation. Following death of the victim,

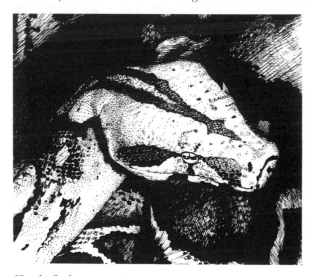

Head of a boa constrictor.

constrictors, like all snakes, swallow their prey whole, opening their mouths widely with jaws attached only by elastic ligaments.

The most well known of the boas is the boa constrictor (*Boa constrictor*). Boa constrictors are common throughout the Neotropics and vary in color pattern. South American boa constrictors are lighter and more sharply marked than their Central American counterparts. They are warm tan with dark brown, diamond-shaped patterning. Their heads, like all boidae, are rather long with pointed snout. Boas average about 5–6 feet in length, but they do not commonly attain the huge lengths sometimes reported. The largest boa constrictor on record was 18.5 feet long, most extraordinary for this species. Boas can be aggressive and will coil, hiss, and bite if attacked. They are mostly nocturnal, feeding on all manner of mammals, including small cats. They also take birds and lizards. Boa constrictors inhabit a wide range of habitats, from wet lowland forests to dry savanna.

The emerald tree boa (*Boa canina*) is probably the most beautiful of Neotropical boids. Deep green above, yellow green below, with a dorsal white line and scattered white spots, it has burning yellow eyes with catlike, slitted pupils. Confined to South America, this 6-foot boa is a tree climber, coiling itself tightly in such a way as to be most cryptic. It has been said to resemble a bunch of bananas when coiled in a tree and, indeed, small individuals have occasionally been found on banana boats among the bunches. The tail is prehensile, and these snakes are skilled at moving about in the trees, preying on squirrels, opossums, birds, and lizards. Both the emerald tree boa and boa constrictor tame when handled frequently by humans.

The rainbow boa (*Epicrates cenchris*) is one of the smaller boas, growing normally to be 3–4 feet long. To be appreciated, this little constrictor must be observed in full sunlight. Its dull blackish brown scales sparkle with iridescent colors, giving the snake its popular name. Rainbow boas are also skilled tree climbers and often prey on bats.

They range from Central America all the way south to Patagonia.

The largest of all the New World snakes is the magnificent golden brown anaconda (*Eunectes murinus*), which ranges throughout Amazonia. Anacondas do not grow to be quite the length of some of the Old World python species, but they can reach 30 feet, though such a length is rare. Anacondas are wider in body than pythons and are considered to be the bulkiest of the world's snakes. One 19-foot specimen, photographed with seven men holding it, weighed 236 pounds. Anacondas are commonest along rivers and marshes, feeding on agoutis, capybaras, peccaries, tapirs, large birds, and even crocodiles and caiman. They do not eat people and will avoid humans by taking shelter under water. Not particularly skilled swimmers, they normally capture prey by lying in wait along quiet muddy marshy riverbanks. Anacondas, like most snakes, are prolific breeders. One recorded birth included seventy-two baby snakes.

Snakes—The Pit Vipers

Pit vipers are all poisonous. Pit vipers (subfamily Crotalinae) range throughout both tropics and temperate zone. North American rattlesnakes, copperhead (*Agkistrodon contortrix*), and cottonmouth water moccasin (*A. piscivorous*) are all pit vipers. The "pits" referred to in the name are sensory depressions located between the nostrils and eyes. They sense heat and aid the snake in locating warm-blooded prey. Pit vipers have long hypodermic fangs in which a poison duct from modified salivary glands can deliver a lethal dose of toxin that attacks both blood and nerve tissue. Pit vipers tend to rest in a coiled position, which they also assume when danger threatens. Many vibrate their tails, and in the rattlesnakes this habit is further enhanced by the presence of the noisy rattles. Pit vipers have large triangular-shaped heads, a helpful feature in recognizing them.

The most infamous Neotropic pit viper is the fer-de-lance (*Bothrops asper*). Known as "yellow-tail" or "yellow-jaw" in some parts of the tropics, this tan snake with dark brown diamond patterning and yellow below can reach 7 feet in length, though most don't exceed 4 feet. Regardless of size, it is a potentially lethal snake: even the juveniles are highly poisonous, and up to fifty young are born at a time. Venom is fast acting and very painful. It rapidly destroys blood cells and vessels and produces extensive necrosis (decomposition) of tissue around the bite site. A person bitten by a fer-de-lance must receive antivenin quickly. Occasionally a fer-de-lance will bite in self-defense but not invenomate. It may have exhausted its venom on a recent catch or simply not discharge it. Though common in lowland forests, the fer-de-lance is abundant also in overgrown tangle, and care should be taken whenever walking through dense jungle. The fer-de-lance feeds on various mammals and some birds. The species ranges widely, from Central America to southern Peru. The name *fer-de-lance* probably refers to the lance-like shape suggested by the long snake body and large triangular head. In Brazil it is called *jararaca*, meaning "arrowhead." It is also commonly called *tomigoff*.

The genus *Bothrops* also contains several species called palm vipers. These snakes tend to grow to shorter lengths than the fer-de-lance but are quite dangerous nevertheless since they are arboreal and usually cryptic, coiled among the branches. They tend to frequent dense jungle and have bitten more than one machete-wielding explorer as he chopped through the underbrush. One species, *B. schlegelii*, is called eyelash viper because of enlarged scales that grow outward over each eye. This species is variable in color, ranging from light bluish to gold. It feeds on small rodents as well as tree frogs and anolis lizards.

Another charming *Bothrops* is the jumping viper (*B. mummifer*), confined to Central America. A short snake with a thick body, gray with black diamonds, this serpent literally hurls itself at its perceived attacker. As it jumps it attempts to bite. The venom is not as potent as that of others in the genus.

A few rattlesnakes, among them the tropical rattlesnake (*Crotalus durissus*), range into the Neotropics. This species occurs from Central America to Brazil and Paraguay usually in dry forests and uplands. Its venom is highly toxic and can cause blindness and paralysis as well as suffocation. It tends often not to rattle the usual warning when approached.

Just as the anaconda is the largest of the constrictors, so the bushmaster (*Lachesis mutus*) is the giant of the pit vipers. Not only is it the largest viper in the Neotropics, it is the world's largest, reaching lengths in excess of 12 feet. It is not a particularly thick serpent compared with others, but its length, large head, and long fangs make it fully as dangerous as a dozen feet of poisonous snake has a right to be. Because of its length, a bushmaster can strike over a long distance. It resembles the fer-de-lance, being light brown with dark brown diamondlike splotches, but its thickest splotches are on its back, not its sides as in the fer-de-lance. The curious Latin name translates to "silent fate," so named for the snake's ability to strike without audible warning (though it usually does vibrate its tail). It is usually considered uncommon, but in some areas, such as the Arima Valley in Trinidad, it is common to find run-over bushmasters on roads at night after heavy rains. Its nocturnal habits probably account for why it is not frequently encountered, and it may, in fact, be frequent in lowland forests. Bushmasters eat mammals and birds.

The Coral Snakes

"Red and yellow, kill a fellow; red and black, friend of Jack." This is a little rhyme to help remember the distinction between North American coral snakes and kingsnakes. Also remember that it does not work in South America. There are fifty species of coral snakes in the New World, and many do not have red bands touching yellow. Its safest just to avoid colorful snakes with red, black, and/or yellow rings. They may be coral snakes.

Coral snakes (genera *Micrurus*, *Micruoides*, and *Leptomi-*

crurus) are members of the global family Elapidae, to which the deadly cobras and mambas belong. Their toxin mostly affects the nervous system and acts very quickly, producing paralysis and death by suffocation. Coral snakes, which reach about 2 feet in length, have short teeth and must chew a bit in order to inject their lethal venom. Their bright patterning is considered to be warning coloration (see chapter 4, page 118). Some nonpoisonous snakes converge in color pattern with coral snakes wherever their ranges overlap, another case of Batesian mimicry (Greene and McDiarmid 1981). When threatened, a coral snake will thrash its body, wave its tail, and try to bite. Handling one is quite dangerous because its narrow head can slip through fingers, giving the snake an opportunity to bite. Coral snakes can be active both day and night but are most often found hiding beneath logs or rocks. You are very unlikely to encounter one but if you do, don't touch it.

Coral snakes occur in habitats ranging from deserts to jungle. They eat mostly lizards and other snakes.

Nonpoisonous Snakes

There are many species of nonpoisonous snakes in the Neotropics. They include the various vine snakes (genus *Oxybelis*), thin brown, gray, or green snakes that climb about the foliage capturing and feeding on lizards. The beautiful indigo snake (*Drymarchon corais*) can reach lengths of 10 feet, eating virtually any kind of animal from fish to birds. The odd large-eyed, chunk-headed snake (*Imantodes cenchoa*) is extraordinarily thin, coiling in outer branches where it preys on small tree frogs and lizards.

Crocodilians

Throughout coastal and inland tropical waters there swim the crocodiles and caymans. These reptiles can grow to lengths well in excess of 10 feet, although smaller-sized animals are more common. The American crocodile (*Crocodylus acutus*) ranges from the Florida keys southward to Ecuador, inhabiting coastal mangrove swamps. In eastern Central America, the Morrelet's crocodile (*C. morreliti*) is an endangered species. Most abundant of the Neotropical crocodilians are the Amazonian caiman species, especially the spectacled caiman (*Caiman crocodilus*) and black cayman (*Melanosuchus niger*). Crocodilians eat fish and other water-dwelling animals, including capybaras, snakes, and birds. After an aquatic mating ritual, females build a nest mound and lay up to sixty eggs. Parent animals, especially the female, aid the newly hatched young in moving from the nest to the water and remain with them for some weeks. Juveniles have many predators, including storks, egrets, raccoons, and anacondas.

Tree Frogs, True Frogs, and Toads

Amphibians are animals that require water to reproduce. Typically, gelatinous eggs are laid in ponds or

Morrelet's crocodile.

streams either as floating masses or attached to rocks or debris. Larval animals hatch and pass through a developmental stage in water (referred to as a tadpole in the case of frogs and toads). During the aquatic phase the animal breathes by external gill tufts, but these are resorbed when the larva passes through metamorphosis to adulthood. Adult amphibians usually require moisture for their skins, though toads are able to survive with dry skins.

Salamanders are not very diverse in the tropics. They represent exceptions to the general tendency for taxonomic groups to show high diversity in the tropics. Some do occur, but by far the most abundant, diverse, and interesting amphibians are the anurans, the frogs, tree frogs, and toads.

Although many species of anurans reproduce in the manner described above, many also show dramatic departures. In Costa Rica, where 119 anuran species are found, N. J. Scott and S. Limerick (1983) have described a dozen different modes of amphibian reproduction. These include live birth, with fully formed miniature adults ("skipping the egg and larval phases"), eggs laid on plants where larvae hatch and drop into water, eggs laid on land in foam nests, in bromeliads, or in tree cavities. Courtship patterns are also sophisticated. Frogs and toads vocalize, the males emitting a specific call that serves to attract the females. Many species are territorial. Amphibian reproductive behavior has undergone an impressive adaptive radiation in the tropics.

Tree frogs are arboreal, aided in attaching to leaves and stems by tiny suction disks on their feet. Most are quite small and well camouflaged, though some are very brightly colored and exhibit warning coloration. One of the most common (and most frequently photographed) is the gaudy leaf frog (*Agalychinis callidryas*). This Central American species has blazing, bulging red eyes, bright green on the upper body along with a scattering of white spots, bluish on the sides, white on the belly, and orange on hands and feet! Males and females have a rather pro-

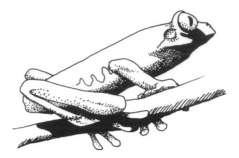

Gaudy leaf frog.

longed mating ritual in which the female, with the male clinging to her, attaches eggs to a leaf as the male fertilizes them. Hatching occurs in approximately five days, and the larvae drop off the leaf into water.

Glass frogs (genus *Centrolenella*) are small green tree frogs with somewhat transparent bellies, revealing the beating heart and intestinal system. They attach eggs to leaves over streams, and larvae hatch and drop into the water. Eggs tend to hatch in heavy rain, facilitating the release of tadpoles into the water below. Tadpoles become bright red and burrow in stagnant litter in slow pools. Their color is the result of a concentrated blood supply, an adaptation to low oxygen levels in the mud.

The most colorful and dangerous of the tree frogs are the Dendrobatidae, the dart-poison frogs. Shiny black, some have bright red or orange markings, some bright green. The Colombian Choco Indians utilize the poisonous alkaloids from the frogs' skins in making potent darts for hunting (Maxson and Myers 1985). The poison affects nerves and muscles, producing paralysis and respiratory failure. These colorful frogs hunt by day, feeding on termites and ants, and it has been suggested that their warning coloration evolved in response to their long feeding periods when they would otherwise be vulnerable to predators. Males make their insectlike vocalizations during the day and attract females. Eggs are laid on land, after the male deposits sperm. The female remains with the fertilized eggs and carries the tadpole to water after it hatches. The female also feeds the developing tadpole highly nutritious unfertilized eggs that she produces. These eggs are deposited following signaling behavior by the tadpole, similar to solicitations of nestling birds when their parent brings food.

Other anurans also contain skin toxins. The giant marine toad (*Bufo marinus*), largest of the New World anurans, secretes irritating fluid from its skin and is quite toxic if eaten. The large size of the animal (easily the size of a softball) would make it a tempting target for predators, but its toxic integument is so dangerous that dogs and cats have reportedly died just from picking up the toad in their mouths. Marine toad tadpoles also have toxic skin. One very curious human application of toad toxin has been studied in Haiti. Along with extract from puffer fish and two plant species, toad toxin is used to induce the deathlike trance observed in victims of voodoo rituals (Davis 1983). The smoky frog (*Leptodactylus pentadactylus*) is another species possessing highly irritating skin secretions. These frogs are aggressive when threatened, inflating their bodies, rising on their hind legs, and hissing.

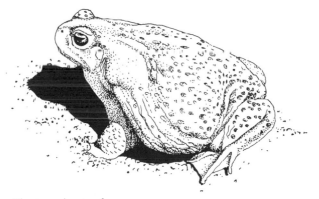

Giant marine toad.

Besides warning coloration and toxic irritating chemicals, frogs have one other defense trick. Some species, which are normally cryptic, display *flash colors* when threatened. A flash color is revealed when the frog raises a foot or other body part revealing a bright patch of color, usually red, orange, yellow, or blue. The sudden bright coloration may momentarily confuse the predator, interrupting its search image long enough for the frog to use its most obvious defense tactic, its jumping ability.

INVERTEBRATES

Myriads of animals without backbones are at home in the Neotropics. From within forest soils to atop the tallest emergent trees, seemingly innumerable creatures crawl, hop, slither, climb, burrow, leap, and fly. In no way can this volume catalog even a representative sample of all the invertebrate life forms that can be found by diligently searching rain forests and other Neotropic habitats. Indeed, new species are found on virtually all collecting trips. Included here is a mere sample of some of the most interesting and frequently seen members of this vast horde. These are some of the invertebrates you are most apt to encounter by casual searching.

The Social Insects

Two great insect orders, Hymenoptera and Isoptera, have evolved species that display complex social systems where societies of close relatives support a fertile queen, their mother. Hymenopterans are bees, wasps, and ants. Not all species are social, but many are. Isopterans are termites, and all are social insects. Both orders are abundantly represented in the world's tropics. Workers are usually separated into differing morphological castes, consisting of various-sized workers and soldiers. Sterile animals are usually females, all sisters, but male workers occur in termites. Otherwise, males are produced only during mating of the queen and are otherwise superfluous. Wilson (1971) provides a comprehensive account of the social insects.

TERMITES

No one can visit the Neotropics without seeing basketball-sized termite nests attached to tree trunks and branches. They occur in all habitats, from rain forest to savanna. My first visit to the tropics, in Belize, began with a drive along a mangrove swamp. I was amazed at the number of termite nests among the red mangroves. From the rounded, blackish brown nests radiate termite-constructed tunnels in which the workers pass to and from the colony. Termites also nest underground in vast subterranean colonies.

The most abundant of the Neotropical termites are the many species of *Nasutitermes*, which range throughout the region (Lubin 1983). These come in three castes: worker, soldier, and queen. If you wish to see workers and soldiers, locate a nest and cut into it. Nests are made of paperlike material called "carton," which is a combination of digested wood and termite fecal material that forms a glue. When a nest is cut into, scores of workers and soldiers swarm out to investigate the disturbance. Workers are pale whitish and lack the constricted abdomens that characterize ants. They have dark mahogany-colored heads. Soldiers are similarly colored but larger than workers with prominent heads and long snouts. Soldiers can eject a sticky substance with the odor of turpentine that apparently irritates would-be predators, including the anteater *Tamandua*. Termites are fast workers and will quickly repair an injured nest. They swarm about the surface laying down new material to replace that which was damaged. Another way to see termites is to break open their tunnels. Workers are blind and follow chemical trails laid down by other workers. They will continue to pass *en masse* along an opened tunnel, though some will eventually repair it. The queen termite (or queens—some species have multiple queens) is located deep within the nest and cannot be seen unless the entire nest is dissected.

Queens are immense compared with workers and soldiers. Virtually immobile, weighed down by a monstrous gourdlike yellow abdomen, their needs are attended to by workers as they pass their lives producing the colony's eggs.

Termites eat wood. They manage to digest the hard cellulose with the aid of a complex intestinal fauna made principally of flagellate protozoans. It is actually the protozoans that digest the cellulose. Removal of these one-celled organisms will prevent a termite from digesting cellulose and other large molecules of wood. It and the protozoans are obligate symbionts, another example of an evolutionary mutualism. The flagellate protozoans benefit by being in the termite in that it provides them with continuous food, shelter, and a means of dispersal, since they go where it does. The very hard woods of many tropical trees are probably in part an evolutionary response to selection pressures posed by termite herbivory. Some woods are apparently toxic to termites.

Termites are so abundant in the world's tropical areas that they may contribute to global climatic warming by enhancing greenhouse effect. According to P. R. Zimmerman and colleagues (1982), their combined digestive abilities produce significant quantities of atmospheric methane, carbon dioxide, and molecular hydrogen. Working in Guatemala, these researchers argued that forest clearance and the coversion of forests to agricultural ecosystems tends to increase termite abundance, thus accelerating the production of the atmospheric gases.

On a more localized basis, Jan Salick and his colleagues (1983) learned that termitaria, the termite nests, form patches of high nutrient concentrations in otherwise nutrient-poor tropical soils. Working along the Rio Negro in Venezuela, they learned that termites consumed between 3 and 5% of the annual litter production, taking it to their termitaria. Termitaria contained more nutrients than litter, and litter was more nutrient-rich than the soil. When termites abandon termitaria, which they do rather regularly (nest rate abandonment averaged 165 nests per hectare per year), these sites form patches of high nutrient level ideal for many tree species. Termite activity has a potentially important influence on nutrient cycling and tree establishment in this area of very poor soils.

ARMY ANTS

The 1954 Hollywood film *The Naked Jungle* portrayed Charlton Heston as a besieged South American plantation owner under attack by billions of army ants. The film, taken from the 1938 Carl Stephenson short story "Leiningen Versus the Ants," portrays the six-legged marauders (called "fiends from hell" in the short story) as moving in a column 10 miles long and 2 miles wide. Nothing could turn the ants from their chosen path, right through Heston's plantation, and the ants ate everything in their path, including attempting to eat Heston's horse, with him on it. Their persistence and ingenuity knew no bounds as they forged streams by making little boats out of leaves. As they approached ever closer to the plantation, the landscape behind them was utterly denuded of all life, a virtual moonscape. Heston survived, but only by flooding his plantation, washing away the formicine horde with diluvian retribution. The portrayal of the ants in the film was ridiculous. Real army ants are orders of magnitude less fearsome than the *mobutu*, as they were called (in hushed tones) by the natives in the film.

South American army ants are nearly sightless (the African ones are totally blind), eat no plants whatsoever, and eat no animals larger than baby birds or small lizards. As I explained in chapter 2 (page 58), should you encounter a column of army ants, simply step over or around them. The only way to be bothered by them is to sleep on the ground in their path (this did happen to someone on one of my trips) or to inadvertently overlook them and step in them. They won't eat you (or your horse), but they will bite, and their bite hurts. Gotwald (1982) provides a comprehensive overview of the world's army ants.

Neotropical army ants are all members of the subfamily Ecitoninae, of which there are 5 genera and 150 species.

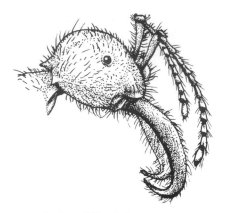

Head of a soldier *Eciton* army ant.

One of the largest and most frequently encountered species is *Eciton burchelli*. *Eciton* varies both in size and color. The largest individuals are soldiers, with extremely large mandibles, sharply hooked inward. The smallest workers are only about one-fifth as large as the soldiers. Color varies from orange and yellowish to dark red, brown, or black depending on subspecies. The easiest way to identify *Eciton* is by seeing soldiers with the huge pincer jaws. Some Amerindians use these formidable mandibles to suture wounds.

Eciton armies are large, often containing over one million recruits. Armies are nomadic, moving through the forest, stopping at temporary bivouacs during their reproductive cycles. They may bivouac only for a night or for several weeks in the same area. When the entire mass moves, it is usually a nocturnal migration. Bivouacs are either underground or in hollow logs or trees and consist of massive clusters of the ants themselves. There is but a single queen, who remains in the bivouac except when the entire army is on the move. Then, like any matriarch, she is attended by an entourage of workers and soldiers.

Each morning the raiders stream from the bivouac, making a dense column in search of prey. Soon the column begins to fan out, raiding parties moving in different directions. Though virtually blind, the ants communicate well by chemical signals, and, once prey is discovered, the ants, being very fast runners, converge on it quickly. This behavior is termed *group predation*, and is somewhat the insect equivalent to a wolf pack (Rettenmeyer 1983). The raiders do not restrict their plundering to the ground. They routinely climb trees, even ascending into the canopy. They will enter human dwellings and poke around the corners and under the boards searching for cockroaches and other small vermin. Smart humans leave the ants to their work. They won't remain long, and the house will be freer of insects for their visit. Captured food is stored temporarily along the marauding columns and later taken back to the bivouac.

Prey consists of anything alive and small enough to subdue, most commonly insects (especially caterpillars), spiders, millipedes, and other denizens of the litter and leaves. Small vertebrates such as tree frogs, salamanders, lizards, and snakes are routinely attacked, and baby birds in the nest are frequent army ant victims. Raiding parties attract much attention. Potential prey attempts to flee, and antbirds and other species gather at the edges of the ant swarm to feed on escaping animals (pages 50 and 259).

THE GIANT TROPICAL ANT

As I mentioned in chapter 2 (page 58), there is a really nasty ant to beware of, the giant tropical ant (*Paraponera clavata*), also called the "bullet ant." This inch-long black ant can both bite and give a very painful sting. Daniel Janzen reports excavating the nest of *Paraponera*, and the ants received him aggressviely. In addition to being stung, he noted a musky odor released as he disturbed the nest, which he hypothesized to have been an alarm chemical. He also heard the worker ants emit a loud squeaking noise (Janzen and Carroll 1983). Not surprisingly, such a formidable ant has its mimics. A beetle species resembles the giant ant when at rest, but the deceptive coleopteran looks like a wasp when in flight. Giant ants tend to be rather solitary and can be found anywhere in the forest, from ground to trees.

Cockroaches

If you manage a trip to the Neotropics without seeing *La Cucaracha* consider yourself to be legally blind. Most of the world's 4000-plus cockroach species live in the tropics, and the New World tropics can certainly lay claim to its share. Most people can identify cockroaches easily, as they routinely cohabit human dwellings, much to the dismay of the bipedal occupants. Cockroaches are oval in shape, the dorsal side covered by a pair of large wings, the head sporting a pair of very long antennae. Excellent fliers and basically nocturnal, they are often seen fluttering around lights at night. They are fast runners and have an uncanny ability to squeeze between floor boards and other tight places.

One of the most striking tropical species is the giant cockroach (*Blaberus giganteus*). This insect behemoth easily fills the palm of your hand and then some. It lives in hollow trees and other reclusive places during the day and scurries about in search of food at night. One favorite food of this and other cockroach species is bat guano. Floors of bat caves literally oscillate, a sea of insects and excrement, as thousands of cockroaches devour the products of bat digestion. In Guatemala, I take my students on a nocturnal field trip. With large powerful flashlights we trek to the outhouses and, in the dead of night, shine the beams down the holes. The cockroach show, mostly *Blaberus*, is something the students talk about long after they return home. Some claim to have even dreamed about it.

Harlequin Beetle

There are many very impressive tropical Coleopterans (beetles), and this, *Acrocinus longimanus*, is one of the finest. It is both large, up to 3-inches, and colorful, being complexly patterned red, black, and yellow, with very long antennae. Males have extremely long and thick front legs, used during mating. Larvae live inside bark, forming galleries inside the wood. Adults live in a variety of trees including figs and are strongly attracted to sap. If you are fortunate enough to find one of these large insects, look carefully under its thick outer wings (elytra). Don't worry, it can't bite you, at least not to the degree that you ought to care about it. Beneath the wings you can usually find a tiny pseudoscorpion (not a real scorpion, so don't worry about being stung). The pseudoscorpion uses the harlequin beetle as a host, an example of commensalism. The host gains no benefit from the pseudoscorpion, but the little hitchhiker does no harm.

Rhinoceros Beetle

The rhinoceros beetle (*Megasoma elephas*) is named for the long upcurved, hornlike projection possessed by the males. A huge and bulky insect, it can reach lengths of over 3 inches. It is a member of the Lamellicornia, a beetle group that includes the scarab and stag beetles. *Megasoma* is a scarab, mostly brownish, its elytra (thick outer wings) covered by tiny hairs. The combination of large size, long horns, and hairy body makes this insect extremely distinctive. If you see one, you'll recognize it. Females are similar to males in size but lack the long horns, a characteristic true of all scarabs.

Darwin (1871) hypothesized that males evolved the

Rhinoceros beetle.

horns by sexual selection, somewhat like the male plumages of the cotingas and manakins I discussed in the previous chapter. The horns, thought Darwin, would aid the males in combat for lovely lady scarabs. Males do use the horns in combat. Scarabs and stag beetles all over the world, from Costa Rica to Africa to the Solomon Islands, jostle like wrestlers, locking horns, the victor lifting the loser and tossing him out of the tree. What is not abundantly clear is the degree to which females care about all this pugilism. Males seem more oriented to fighting for favored feeding sites in the trees than for females (Otte 1980). Females may "care" about the battles insofar as victorious males have access to good sources of nutrition. From a female's viewpoint, these would be the males to get to know. Nonetheless, in many species, males have short horns (Brown and Rockwood 1986), and the adaptive significance of beetle horns remains to be solved.

There are many scarab species in the Neotropics, most very beautiful. This, of course, has led to their downfall since beetle collectors relish them. In addition, species such as *Megasoma* require mature lowland rain forest because the larvae must live in large decaying logs. The cutting of the rain forests and conversion from forest to brushy areas may significantly reduce the beetle's reproductive success (Howden 1983). Already, *Megasoma* is considered to be a rare species.

Lantern Fly

This remarkable insect (*Fulgora laternaria*) is really worth seeing. It is large, with a 5-inch wingspan, and when oriented vertically along a tree trunk, it vaguely resembles a lizard. The reason for the resemblance is its long head, with markings reminiscent of a cross between the heads of a lizard and an alligator. Indeed, one Spanish name for it is *mariposa caiman* meaning "alligator-butterfly." A member of the Homoptera, or sucking insects, the lantern fly is unusual in almost every way, including its contribution to Neotropical folklore. Its English name,

Lantern fly.

"lantern fly," is derived from the mistaken belief that its huge head is bioluminescent, glowing in the dark. It isn't and it doesn't. Another even more startling piece of folklore is that if a young girl is stung by a lantern fly (or *machaca*, as the locals call them), she must have sex with her boyfriend within twenty-four hours or she will die from the bite (Janzen and Hogue 1983). This legend was probably authored by a creative but anonymous young Indian man from centuries past who kept company with a coy and perhaps gullible young Indian woman.

The lantern fly comes equipped with several survival strategies (Janzen and Hogue 1983). When oriented on a tree trunk, it is highly cryptic, its soft mottled grays making it appear as part of the bark. If disturbed, it will climb away or drum its head against the tree making a rapping sound. If the disturbance persists, the insect discharges a skunklike odor and flies to another tree. When it takes flight, it reveals bright yellow eye spots on its hind wings, rather like those of the owl butterfly (see below). These spots, which are also revealed by a quick flash of the wings, may act to temporarily confuse a would-be predator.

Selected Butterflies and Moths

There are some butterflies that most every visitor to the lowland rain forests notices. I've previously discussed *Heliconius*, its relationship with *Passiflora*, and its mimicry complexes (chapter 5). Now I'll briefly introduce you to a few more butterflies and one moth. DeVries (1987) provides a complete guide to Costa Rican butterflies.

Put out a few overripe bananas at night, and soon both a large brownish butterfly and moth will appear. The owl butterfly (*Caligo memnon*) and the black witch (*Ascalapha odorata*) are both among the largest of the tropical lepidopterans, and they tend to be strongly crepuscular, active at dusk and dawn, though adults fly during the day in deep rain forest shade. They feed on rotting fruits, hence their orientation to bananas. Black witches are occasionally mistaken for bats in the evening twilight as they flutter about a banana bunch, and this species is frequently flushed from shady sites such as hollow trees during the day, again giving rise to the mistaken notion that it is a bat.

Like the lantern fly, the owl butterfly has a prominent eye spot on the base of each wing. The spot is on the underside of the wing and is commonly flashed when the otherwise cryptic animal is disturbed on its tree trunk resting place. Only one spot is flashed, not two, although both wings have spots. Many individuals show damage from bird beaks in the vicinity of the eye spot giving rise to the notion that the spot "redirects" the bird's attack allowing the butterfly to escape with little damage. Another suggestion is that the eye spot serves to mimic a distaste-

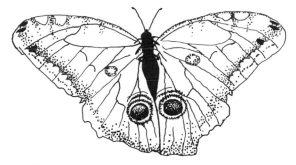

Owl butterfly.

ful tree frog (Strandling, cited in DeVries 1983). The upper side of the wings is soft bluish, somewhat similar to the blue morpho (see below).

The blue morphos (genus *Morpho*) are every butterfly collector's dream. Large, with brilliant deep blue upper wings that seem to glow in sunlight, morphos are deceptively swift fliers able to elude the most persistant wielder of an insect net. Particularly common around streams and other sunlit areas, they feed on a wide variety of plant species. When on the ground or on a tree, morphos are cryptic, but they are obvious in the air. Their striking patterning and color, visible only in flight, has been termed the "flash and dazzle" strategy of capture avoidance (Young 1971). There are about eighty species of blue morphos, and all are members of the family Morphidae.

There is a group of butterfly species that accompanies army ant swarms, especially *Eciton burchelli* (see above). These so-called antbutterflies are all members of the large family Nymphalidae and subfamily Ithomiinae, and their collective common name is "army ant butterfly." They generally resemble *Heliconius* and are part of the large "tiger-stripe" mimicry complex (page 207) that includes heliconiines, danaines, satyrids, other ithomiines, and a moth or two. Only female army ant butterflies actually orient to the army ant swarms. Thomas S. Ray and Catherine C. Andrews (1980) noted anywhere from eight to twelve flying about the swarm, and they captured up to thirty within a few hours. The butterflies were feeding on the droppings of ant-following birds. Females were probably using the droppings as a nitrogen source, necessary in making eggs.

Often along the roadsides there will be aggregations of butterflies surrounding a stagnant pool. More times than not, these butterflies will be bright yellow and orange, members of the genus *Phoebis*, otherwise called the sulphurs. Males are bright yellow with orange on the forepart of the wing, and females are more uniformly orange with black lining the outer part of the wings. Sulphurs range from the United States through the Neotropics and are inhabitants of open areas and forest edges. They feed on many of the flowers, such as *Lantana*, which are the common tropical roadside weeds. They aggregate at pools containing urine from cattle or humans. The urine supplies them with sodium and nitrogen, just as bird droppings do for army ant butterflies.

Forest-floor Millipede

This species, *Nyssodesmus python*, is a common denizen of the leaf litter. Females attain a length of about 3.5–4.0 inches, and males are a bit smaller. They look well armored with a flattened shiny carapace protecting their delicate undersides. Colorful, the little beast is yellowish with two orange-brown stripes down its sides.

Millipedes are harmless ambling herbivores and should not be confused with swift moving carnivorous centipedes (chapter 2, page 50). When threatened, millipedes roll up in a little ball and tough it out. They also have an impressive array of chemicals at their command. The hindgut can squirt a volley of noxious liquid, containing both hydrogen cyanide and benzaldehyde, up to a foot away (Heisler 1983).

PERIPATUS

A *Peripatus* is a most unusual wormlike animal with characteristics of both annelids (segmented worms) and

Forest-floor millipede.

arthropods (joint-legged animals). For this reason, *Peripatus* is usually classified in its own phylum, Onychophora. About thirty species of Onychophorans crawl about in the world's tropics, including Australia, New Zealand, New Guinea, Malay Peninsula, Africa, and the Neotropics. Though segmented like an annelid, *Peripatus* has the beginnings of an exoskeleton, a basic arthropod characteristic. Its stubby, unjointed legs terminate in claws, another arthropod characteristic. Eyes are simple, not compound.

Found among the denizens of forest floor litter, *Peripatus* ambles along on short unjointed legs, resembling a slender caterpillar with stubby, segmented antennae. It captures insects by shooting them with a gluelike substance that engulfs prey in a network of sticky threads. This odd adaptation also helps defend *Peripatus* from would-be predators (Ghiselin 1985).

CHAPTER 8

Tropical Savannas

ALTHOUGH most people conceive of the Neotropics as rain forest, those who have actually traveled in Central and South America know differently. There are vast grassland ecosystems, some wet, some dry, some with scattered trees, some with scattered shrubs, located throughout tropical areas, collectively termed *savannas*. J. S. Beard (1953) was among the first researchers to define tropical savanna, calling it "a natural and stable ecosystem occurring under a tropical climate, having a relatively continuous layer of xeromorphic grasses and sedges, and often with a discontinuous layer of low trees and shrubs." In Beard's definition, "xeromorphic" refers to plants adapted to withstand periodic dryness, and the key element of the definition is that savannas are stable. They do not succeed to forest or any other ecosystem type.

The tree species that populate savannas throughout Central America, the Caribbean islands, and equatorial South America are palmettos, palms, cecropias, and others, depending upon location. Local plant species composition varies considerably. In much of Central America the most abundant savanna tree species is Caribbean pine (*Pinus caribaea*), often adorned with bromeliads and orchids. Several species of oaks are common in Central American savannas, though no oaks are found in South American savannas. Fire-resistant tree species such as *Byrsonima crassifolia*, *Casearia sylvestris*, and *Curatella americana* are abundant on South American savannas. Grasses and (in wetter areas) sedges form the ground vegetation. Soil, though it can vary widely, is either sandy or claylike, typically being described as "poor soil."

Throughout parts of northern South America, particularly within the Orinoco Basin, are highly seasonal, often wet savannas called *llanos*, where sedges and grasses

Savanna.

dominate and trees and shrubs are widely scattered, often occurring as "island" woodlots called *matas* (Blydenstein 1967: Walter 1973). Llanos resemble the Florida everglades, being essentially tall grass prairies with grasses growing up to 3 feet in height. Depending upon season and elevation, llanos may be either wet or dry. Some llanos are quite wet and support stands of Moriche palm (*Mauritia* sp.).

The largest area of savanna vegetation occurs in central Brazil, the *campos cerrados* forming a wide belt across the country from northeast to southwest (Sarmiento 1983). Cerrados occur on acidic, deep, sandy soil. Vegetation type ranges from open woodlands with a 4- to 7-meter canopy to dense scrub thicket (Whittaker 1975). Cerrado soils are nutrient-poor, a factor probably partly responsible (see below) for their existence, rather than richer forest. Crop yields are dramatically increased when soil is fertilized with trace elements (Walter 1973).

In extreme southern Brazil continuing southward through Patagonia is an area termed *pampas*. Mostly grassland, the pampas are not considered to be true savanna, which is confined to the tropics and subtropics, but are part of the southern temperate zone. Biogeographers distinguish between the term *savanna*, for tropical and semitropical grassland, and *steppe*, for nontropical grassland. The pampas are steppes, and consist of vast stands of tussock grasses (*Stipa brachychaeta*, *S. trichotoma*). In areas of sandy soil and decreased rainfall, dry woodland occurs, consisting mostly of a single species, *Prosopis caldenia*. Some very unique animal species inhabit pampas, and I will briefly describe some of the most interesting later in the chapter.

What Causes a Savanna?

There is no simple or single environmental factor that determines that a given site will be savanna (Bouliere and Hadley 1970; Huber 1982, 1987). G. Sarmiento and M. Monasterio (1975) put the paradoxical nature of savannas well: "It is remarkable that savanna is the only ecosystem in the entire warm tropical region whose origin and permanence have been considered unanimously as a fundamental ecological problem." Neotropical savannas occur on a wide variety of soil types, experience all extremes of tropical climate; rainfall may be seasonal or nonseasonal; and water drainage may be rapid or slow (Huber 1987). Fire is common, and savannas tolerate fire well, rebounding quickly after burning.

All savannas occur at low altitudes, less than 1200 meters above sea level. Most savannas are located in environments where the amount of moisture is somewhere between what characterizes rain forest and dry deciduous or thorn forest. In other words, on a scale from very wet to very dry, savannas usually are in the midrange. This suggests that climate has a strong influence on savanna formation, but it cannot be the only influence because often savanna occurs in the midst of otherwise wet forest areas. For this reason, local soil type must also contribute to savanna formation. Soil and climate can interact in dramatic ways. In the central llanos, heavy rains interact with soil to form a hardened crust of lateritic ferric hydroxide, usually at some depth in the soil but occasionally on the surface. This crust, termed *Arecife*, is sufficiently hard to impede the growth of tree roots, except where the woody species encounter channels through the crust. Tree groves or *matas* occur where roots have penetrated Arecife, resulting in the clustering of trees. Grass, with shallower root systems, usually thrives above the level of Arecife (Walter 1973).

Fire is believed to be an important influence on both savanna formation and propagation. Natural fires set by lightning are common, especially in areas with a pronounced dry season. Some savannas form on sites where rain forest has been repeatedly cut and burned, suggesting that human activity can change the site from forest to savanna.

Below, I review each of the major factors determined to be important in savanna formation.

Climate

Savannas typically experience a rather prolonged dry season. One theory behind savanna formation is that wet forest species are unable to withstand the dry season, and thus savanna, rather than rain forest, is favored on the site. Savannas experience an annual rainfall of between 1000 and 2000 millimeters, most of it falling in a five to eight month wet season. Though plenty of rain may fall on a savanna during the year, for at least part of the year, little does, creating the drought stress ultimately favoring grasses. Such conditions prevail in savannas throughout Venezuela, Colombia, Bolivia, Surinam, Brazil, and Cuba, but many savannas in Central America (Nicaragua, Honduras, Belize) as well as coastal areas of Brazil and the island of Trinidad do not fit this pattern. In these areas, rainfall per month exceeds that in the above definition. For only three months at the most is rainfall below 100 millimeters per month. Other factors must contribute to savanna formation in these areas.

Soil Characteristics

Some savannas occur on wet water-logged soils, others on dry, sandy well-drained soils. This may seem contradictory, but it only means that *extreme* soil conditions, either too wet or too dry for forests, are satisfactory for savannas. More moderate soil conditions support moist forests.

Waterlogged soils occur in areas of flat topography or poor drainage. These soils usually contain large amounts of clay and easily become water-saturated. Air cannot penetrate between the soil particles, making the soil oxygen-poor. In extreme cases, hardened pans form, as in the case of Arecife cited above.

By contrast, dry soils are sandy and porous, their coarse textures permitting water to drain rapidly. Sandy soils are prone to the leaching of nutrients and minerals (page 72) and so tend to be nutritionally poor. Though most savannas are found on sites with poor soils (either because of moisture conditions or nutrient levels or both), poor soils can and do support lush rain forest. The white, sandy soils of the upper Amazon (page 73) support such forests, unless the forest is cut and burned (see below).

Fire

The vast savannas of Nicaragua, Honduras, and Belize are populated abundantly by Caribbean pine. Riding through miles of savanna along the Southern Highway in Belize, one notices that many of the pines have darkened fire scars on their trunks. Indeed, midway between the coastal towns of Dangriga and Punta Gorda, is located a fire tower affording a magnificent panorama of savanna. For most of the year, a pair of barn owls (*Tyto alba*) are the only inhabitants of the tower. However, during the dry months of the spring, the tower is manned by someone on the lookout for fires. Lightning storms commonly set fires during dry season, and the effects of dryness and periodic fires combine to preserve savanna. Caribbean pines tolerate occasional mild fires better than other tree species in Belize. Grasses also thrive in an environment with periodic fire.

Most savannas probably experience mild fires frequently and major burns every two years or so. Savanna species are called *pyrophytes*, meaning they are adapted in various ways to withstand occasional burning. *Frequent* fire is a factor to which rain forest species seem unable to adapt, though ancient charcoal remains from Amazon forest soils dated prior to human invasion suggest that occasionally moist forests also burn. If fire did not occur in savannas, trees and shrubs associated with rain forests would tend eventually to replace savanna species and savanna would gradually change to forest. Many successional genera typical of rain forests (*Cecropia, Heliconia, Piper,* and *Miconia*) grow on Belizean savannas, forming thickets very similar to rain forest successional sites. I have little doubt that fire is a crucial determinant of savanna, at least in Belize.

Human Influence

On certain sites, particularly in northern South America, savanna formation seems related to frequent cutting and burning of moist forests by humans. Increase in pastureland and subsequent overgrazing has resulted in an expansion of savanna. The thin, upper layer of humus is destroyed by cutting and burning. Humus is necessary for rapid decomposition of leaves by bacteria and fungi and recycling by surface roots (page 73). Once the humus layer disappears, nutrients cannot be recycled and leach from the soil, converting soil from fertile to infertile and making it suitable only for savanna vegetation. Forests on white, sandy soil are most susceptible to permanent alteration.

Not all rain forest sites are convertible into permanent savanna. As discussed in chapter 3, some rain forests are located on young, rich soils, and these sites will continue to succeed back to rain forest following cutting. It is moist forests on poor lateritic soils that are in jeopardy if cut repeatedly. Once savanna takes over such a site, rain forest may never reoccupy it.

Are Savannas "Natural"?

Some ecologists have suggested that virtually all Neotropical savannas have resulted historically from human activity. This claim is unsubstantiated by historical evidence, however. Analysis of preserved pollen suggests that savannas were present long before people arrived and, thus, are a naturally occurring ecosystem type in the region. Evidence exists that savanna vegetation grew in the Amazon Basin as recently as 13,000 to 30,000 years ago (Sarmiento and Monasterio 1975). Jürgen Haffer (1969, 1974) has provided evidence of the importance of expansion and contraction of savannas during the Pleistocene in determining many present-day patterns of species diversity (page 151).

Though savannas in general contain far fewer species than rain forests, Neotropical savannas demonstrate the highest plant species richness of any savanna ecosystems on the planet (Huber 1987). In numbers of both herbaceous and woody species, Neotropical savannas rank first. Such a high species richness suggests that evolution of savanna species has been occurring for many millennia prior to human arrival and that savanna is as unique and intrinsic to the Neotropics as rain forest.

The origins of savanna clearly vary regionally. Sarmiento and Monasterio (1975) summarized savanna causation, concluding that three basic savanna types exist.

The *nonseasonal savanna* is largely the result of poor soils. It is the savanna of white, sandy soils, where drainage is rapid and climate is wet most of the year.

The *seasonal savanna* is the most widespread type in the tropics. It occurs on sites with a stressful dry season where soils are sandy and nutrient-poor. Fire is an important component of these savannas.

The *hyperseasonal savanna* is characterized by an annual

period of water deficiency plus a period of saturation. In other words, there is either too much or too little water in the soil, making soils nutrient poor and tending to waterlog. These savannas are typically all grass, with very few trees and shrubs.

There is a dynamic, temporal interface between savannas and moist forests. One expands while the other contracts in a climatically driven, edaphically influenced, long-term equilibrium that has produced and continues to produce far-reaching effects on evolutionary patterns of both plants and animals.

WHAT LIVES IN SAVANNAS?

Species richness of savannas is much less than rain forests. Expansion of savanna either by climate or by human alteration of the landscape will contribute to loss of rain forest species. Mammals such as monkeys, sloths, tapirs, and peccaries are either absent or highly uncommon in savannas. On the other hand, deer, gray fox, jaguarundi, tayra, and most armadillos inhabit savannas as well as forests. Anteaters venture into savannas in search of their formicine prey, as do the 5-foot long giant armadillos (*Priodontes giganteus*). These creatures may attain weights of up to 130 pounds. The giant anteater (*Myrmecophaga tridactyla*) is more common in savannas than rain forests. Many rain forest species venture into savannas that abut forest but will not extend their ranges far into savanna.

Savannas are ideal habitats in which to observe birds. The llanos, vast, wet grasslands support a diverse assemblage of waterbirds. James A. Kushlan and two colleagues (1985) studied the wading bird community of Venezuelan llanos and found twenty-two species of large wading birds, including seven ibis species, one spoonbill, eleven herons and egrets, and three storks. They compared the llanos with the Florida everglades, where only fifteen species of large waders occur. Herons were the most diverse species in both ecosystems. Stork species, including the wood stork (*Mycteria americana*), maguari stork (*Euxenura maguari*), and the huge jabirou (*Jabiru mycteria*) were richest on the llanos, probably because large fish, their principal prey items, were more abundant than in the everglades. Only the wood stork occurs in the everglades. The researchers noted differences among species relative to their foraging behaviors, prey selectivity, and habitat use. They hypothesized that the greater diversity of waders on the llanos is due in part to greater habitat diversity, increasing the types of feeding areas available. They noted that several species, including the buff-necked ibis (*Theristicus caudatus*) and sharp-tailed ibis (*Cercibis oxycerca*), fed in very shallow habitats on high ground. Such areas were less available in the everglades. They also noted that during the dry season many llanos species fed together, using similar foraging behaviors and feeding sites. They speculated that prey availability is so high that competition among the wader species is minimal.

Three vulture species, the black (*Coragyps atratus*), turkey (*Cathartes aura*), and king (*Sarcoamphus papa*) can often be seen in the sky together over savannas. The black vulture is one of the most ubiquitous birds of the Neotropics, ranging from Argentina well into North America. Black vultures commonly congregate in vast numbers around garbage dumps and are thus common city dwellers. Turkey vultures are named for their red heads, giving them a superficial resemblance to turkeys. Turkey vultures fly with wings distinctly upraised, in a dihedral pattern. One hawk, the zone-tailed (*Buteo albonotatus*), flies very much like a turkey vulture, a possible form of behavioral mimicry, since the plumage of the two species is generally similar. Perhaps a zone-tailed hawk can fly closer to potential prey if it's mistaken for a carrion-eating turkey vulture. The king is the largest and most spectacularly plumaged Neotropical vulture. It is black and white, and its head is adorned with bright (but bizarre-looking) orange wattles.

Many raptor species frequent savannas, searching for prey in the open habitat. The crested caracara (*Polyborus plancus*) with its ragged black crest, can often be observed walking about on the ground. Like vultures, it is a carrion feeder. The rufous-colored savanna hawk (*Heterospizias meridionalis*) is also often on the ground especially after fires, when it searches for small mammals, reptiles, and insects driven out by the flames.

Passerines provide interesting bird-watching in savannas. The gray and black fork-tailed flycatcher (*Muscivora tyrannus*) performs aerial acrobatics in quest of insects. Adult fork-tails have streamerlike tails twice the length of their bodies. The brilliant red and black vermilion flycatcher (*Pyrocephalus rubinus*) conspicuously dashes from a perch in pursuit of some luckless insect. Some of the many members of the huge ovenbird family (page 257) nest in savannas, including the unique-sounding pale-legged hornero (*Furnarius leucopus*). This bright rufous-colored bird has a distinct, loud song made up of a series of high-pitched notes that go downscale. The hornero constructs a large mud nest shaped like a dome.

Thomas R. Howell (1971) did extensive research on the bird community of a lowland pine savanna in northeastern Nicaragua. He found only 56 species in savanna compared with 116 species in nearby rain forest area. Of the savanna species, 26 were permanent residents, including such species as chipping sparrow (*Spizella passerina*), hepatic tanager (*Piranga flava*), eastern meadowlark (*Sturnella magna*), and red crossbill (*Loxia curvirostra*). These birds also occur in similar habitats in North America. Howell described them as savanna specialists, species that could not possibly colonize rain forest. Howell noted that the diversity gradient (page 122) between the temperate and tropic zones is not present when comparing savanna bird communities. A savanna in Georgia contained exactly the same number of nesting species (15) as that in Nicaragua and far more individuals. Toucans, antbirds, manakins, cotingas, motmots, and trogons were all common in rain forest but absent from savanna. Only hawks and vultures were significantly more diverse on savanna. Nineteen bird species that occurred on savanna were ac-

Laughing falcon, a common savanna
predator, specializing in snakes.

tually dependent on broad-leaved forest and only utilized
savanna for occasional feeding.

Snakes and lizards of various species occur commonly
in savannas. I encountered my first boa constrictor as it
crossed a savanna road in Belize.

Scorpions and tarantulas also frequent savannas, and
certain insects, like tiny biting flies can be an extreme nuisance.

A Quick Look at the Pampas

The dry, barren pampas host a variety of animals not
characteristic of rain forests. The pampas deer (*Ozotoceras
bezoarticus*) is endemic to the pampas as is the maned wolf
(*Chrysocyon brachyurus*). Rodents are abundant, including

the mara (*Dolichotis patagonum*), sometimes called the patagonian hare. Superficially resembling a hare (including long ears), this rodent, also called the patagonian cavy, can leap 6 feet.

Charles Darwin was fascinated by a group of burrowing rodents of the pampas collectively called the tucotucos (*Ctenomys* spp.). Writing in *The Voyage of the Beagle*, Darwin said, "This animal is universally known by a very peculiar noise which it makes when beneath the ground. A person, the first time he hears it, is much surprised; for it is not easy to tell whence it comes, nor is it possible to guess what kind of creature utters it. The noise consists in a short, but not rough, nasal grunt, which is monotonously repeated about four times in quick succession; the name Tucutuco is given in imitation of the sound."

The largest and perhaps most spectacular birds of the pampas are the flightless rheas, relatives of the ostrich. Two species, the common rhea (*Rhea americana*) and Dar-

Rhea.

win's rhea (*Pterocnemia pennata*) occur both on pampas and more northern savannas. The common rhea is the larger and more abundant of the two. Rheas have the unusual habit of laying eggs in a communal nest. Several females mate with one male, and each hen deposits two to three eggs in the same nest. Only the male incubates. Rheas run very swiftly but were successfully hunted by gauchos, the horsemen of the pampas, who brought them down using the bola, a twine tied to three balls, which was skillfully tossed from horseback to entangle the bird's neck or legs.

CHAPTER 9

Coastal Ecosystems: Mangroves, Seagrass, and Coral Reef

MANGROVES

Forests of mangroves line tropical coasts, lagoons, and offshore islands (called cayes). The various mangrove species are short trees, all of which are highly tolerant of immersion in salt water. These trees colonize shallow sandflats, trapping sediment and gradually building a dense, muddy, organic soil. Mangroves are effective colonizing species and line tropical coasts. They rebound well after disturbance (such as a hurricane). Mangroves provide important nesting sites for birds as well as shelter and food for many fish and invertebrates. Mangrove ecology is summarized in Walsh (1974), Lugo and Snedaker (1974), and Rodriguez (1987).

Types of Mangroves

Mangrove is a West Indian term referring to a group of short trees and shrubs inhabiting tropical coasts. Unlike the names *oak* or *maple*, both of which refer to a specific genus of closely related species, mangrove is a general term for any salt-tolerant tree species of the tropics. For this reason, there is not total agreement among botanists as to exactly which species deserve the term mangrove. One classification has noted twelve genera of mangroves representing eight families, while another cited eighteen genera in nine families. However, there is total agreement that the four species I discuss below are legitimate mangroves. These are the most common mangroves of the Neotropics.

All mangroves are woody plants that tolerate high internal concentrations of salt, a consequence of immersion in high salinity tropical seas. These unique plants have salt glands on their leaves that remove excess sodium and chloride. Mangroves are also tolerant of soils low in oxygen. The thick, muddy substrate that anchors them is virtually anaerobic: devoid of any gaseous oxygen. Mangrove leaves are similar to leaves of many rain forest tree species in that they are simple and unlobed and very thick with a heavy waxy cuticle. This aids in storing water and preventing excess water loss through transpiration. Their major control of gas exchange is through stomatal openings on the leaves.

Red mangrove (*Rhizophora mangle*) is easily recognized. This abundant species can grow as a bushy shrub or a 30-foot-tall tree. It has reddish bark and numerous aerial prop roots, some of which are firmly anchored to the substrate and some of which grow downward toward it. Prop roots are important in providing a firm anchor for the

Red mangrove. Note prop roots anchoring plant in soft mud-sand.

plant since red mangrove is often a pioneer species, colonizing newly exposed areas. The broadly spreading roots provide stability against winds, tides, and shifting sands. Prop roots also contain openings called lenticels, important in transporting air to the oxygen-starved deep roots. Leaves are oval and thick, dark green above, yellowish below. Flowers are pale yellow with four petals. Fruits are reddish brown and produce elongate green seeds that actually germinate while still attached to the parent plant. Seedlings resemble green pods and are about the length of a pencil. They drop from the plant and float horizontally in the sea, becoming flotsam in the tropical ocean. Dispersal is effective, and red mangrove has populated all tropical seas around the world. Seedlings eventually absorb sufficient water that they orient vertically as they float in the sea. Should the tide carry the vertical seedling to a shallow area, once it touches substrate it will anchor and begin to put out roots. Red mangrove ecology is discussed by Golley et al. (1962) and Onuf et al. (1977).

Another easily recognized mangrove is black mangrove (*Avicennia germinans*). Black mangroves tend to grow in less exposed areas than reds. They are bushy topped trees that reach heights of 60–70 feet. Leaves are oval, leathery, and downy white underneath. The flower is yellow and tubular, the fruit green and oval. Seedlings float and are highly tolerant of low oxygen levels. The most notable feature of black mangrove is its root system. Shallow horizontal roots anchor it in thick anaerobic mud, but these roots send up vertical shoots, above ground, called pneumatophores. Lenticels on the pneumatophores feed into wide air passages connecting with underground roots, providing a means for air transport to the oxygen starved root system.

Two other common mangroves are the white mangrove (*Laguncularia racemosa*) and buttonwood (*Conocarpus erecta*). White mangrove tends to grow at slightly higher elevations than red and black mangrove. It is less tolerant of prolonged immersion in the sea. White mangrove has scaly reddish brown bark and greenish white

Inside a black mangrove stand. Note pneumatophore roots protruding up from mud.

flowers. It grows from 30 to 60 feet tall. Buttonwood resembles white mangrove but occurs only well away from daily flooding by salt water. It is the least salt-tolerant of any of the four common mangroves. It was once not even considered to be a mangrove.

A Walk Through a Mangrove Caye

Man-o'-War Caye, 10 miles east of the Belize mainland, is typical of most of the mangrove cayes throughout the American tropics. There is a bird colony located on it, mostly composed of magnificent frigatebird, or man-o'-war birds (*Fregata maginificens*). It is the birds that give the caye its name. Our sailboat nears the island, passing over shallow beds of seagrass (see below) interspersed with bare areas of white coral sand. A spotted eagle ray (*Aetobatus narinari*) glides along ahead of the boat clearly visible in the transparent Caribbean Sea, and black long-spined sea urchins (*Diadema antillarum*) dot the bottom. Brown pelicans (*Pelecanus occidentalis*) dive for fish in the shallow water. It is a hot, sunny January day, and most of the frigatebirds are airborne, gliding like immense slender kites against the deep blue sky.

Our attention is drawn to the mass of birds above us. Frigatebirds are slender, glossy black, with a forked tail and 7-foot wing span. They use their sharply hooked bills to snatch fish from the water surface. They also pursue and rob gulls, pelicans, and boobies of their prey, a behavior that earned them the name "man-o'-war bird." Females have black heads and white breasts, and juveniles have all white heads and breasts. The breeding males have bright scarlet throat sacs that, when inflated, dangle loosely like balloons from their necks. From a distance, the inflated red throats resemble so many Christmas balls dotting the mangroves. As the boat prepares to land on the caye, the birds become increasingly nervous, clattering their bills and flying from their flimsy stick nests. A group of about a dozen brown boobies (*Sula leucogaster*), smaller than the frigatebirds, with brown head and body and white breasts, leave their nests and fly past us. One male frigatebird almost touches the mast with his wings as he flies off, his huge throat sac swinging awkwardly from side to side. The rank odor of the guano, the accumulated excrement on the leaves and mud, is a reminder of the role of these birds in concentrating nutrients. The mangroves are dense on this caye because the bird guano helps fertilize the soil.

The boat ties to some red mangrove prop roots that line the edges of the caye. We must climb under and around the matrix of roots in order to set foot in the slip-

Magnificent frigatebird males.

pery, odorous mud that is the caye. White sneakers turn black as they encounter mangrove mud for the first time. Though the base of the caye is coral sand, the sand is thoroughly glued together by the sticky mud. Above us, birds still perched look down as we notice dead chicks fallen from their nests, decomposing as they hang limply among the red mangrove branches. Here and there is an adult that suffered the same fate. Landing in mangroves is not all that easy for such large birds.

The mangrove roots are anything but barren of life. Among them are numerous mangrove tree crabs (*Aratus pisonii*) and land hermit crabs (*Coenobita clypeatus*), the lat-

Brown pelican.

for sporting all manner of snail shells. Mostly active at night, some land hermit crabs burrow, and others tuck in for the day among the mangrove roots. They have one very large strong claw and do not hesitate to use it, as they do not like being disturbed. Also on board the trees are mangrove periwinkles (*Littorina angulifera*), small snails that cluster about the branches. Here and there among the crabs and periwinkles is a tarantula. The mud, like the tree branches, is also heavily populated by crabs, in this case mangrove fiddler crabs (*Macropipus puber*). As we try to walk through the morass of mud and roots, these tiny crabs scurry to their innumerable burrows. The mud

Land hermit crab.

seems alive with them. One would not like to contemplate being staked out here at night.

Once inside the line of red mangroves we are among black mangroves. Small pools of stagnant water are interspersed among the mud, and the entire area is peppered with pneumatophores from black mangrove roots, poking up from the mud. Walking over these stubby vertical roots requires some getting used to. An amazing array of flotsam has been trapped, including glass and plastic bottles, tooth brushes, hunks of plywood, an old hat, somebody's cracked sunglasses. These are but a few of the objects washed into the caye, trapped by the mangrove roots and mud. Looking at the junk, its easy to see how these cayes grow by trapping sediment.

In the innermost part of the caye, where sediment is thickest, we locate a clump of 30-foot-tall white mangroves. These trees are taller than red mangroves, though a few black mangroves are about equal in size. Singing from the branches is a mangrove warbler (*Dendroica erithacorides*), now considered by some to be a subspecies of the widely spread yellow warbler (*D. petechia*). It is a small slender bird with a dull red head and yellow body. Nearby is an iguana. How it got to the caye is any-body's guess.

How a Mangrove Caye Develops

Red mangrove lines the outer edge of the caye, and, in the sea just beyond the caye, red mangrove saplings grow. A careful look at the pattern of mangrove distribution in the caye strongly suggests a zonation among the various species. Outermost are "pioneer" red mangroves, fol-lowed by black mangroves, and innermost are white man-groves and an occasional buttonwood. This zonation pat-tern correlates generally with the tolerance each species has for salt water immersion. Reds are most tolerant, whites and buttonwoods much less so. Further, the caye appears to be expanding outward. Red mangroves con-tinue to colonize the outer edges of the caye, but as sedi-ment builds and the caye rises, black mangroves expand their ranges outward, essentially moving in on the reds, pushing them further outward. Whites and buttonwoods likewise expand their ranges as the sediment builds. Such a pattern has also been described for mainland coastal mangroves. Reds are outermost, blacks intermediate, whites and buttonwoods innermost.

Many authorities dispute the idea that mangroves are sharply zoned and represent a sort of successional se-quence (Rodriguez 1987). Not all mangrove areas seem to accumulate sediment, and changing conditions caused by storms and tides could influence the pattern, making it difficult to determine competitive relationships among the species. Some mangrove cayes have remained stable for many years, without significantly expanding or con-tracting.

Nearby Man-o'-War Caye is Cocoa Plum Caye. It has borne the full brunt of several hurricanes, being severed at least twice. What was one island became two, which, after another storm, became three. Cocoa Plum bears the scars of its storms. Uprooted mangroves, dead mangrove stumps, and wide open sandy areas make for picturesque

photographs. However, among the fallen and dead mangroves are numerous seedlings and saplings of red and
black mangroves. Without a doubt, the caye is being recolonized and will be rebuilt by the trees. Around the
caye are shallow areas of submerged blackened peat,
which were once part of the island. These areas illustrate
the dynamic nature of a mangrove island. Storms change
its boundries, but the mangroves persist.

SEAGRASS

Closely associated with the mangrove cayes are the vast
flats of seagrass that populate the clear shallow sea inside
coral reefs. Seagrasses are flowering plants, not algae, but
they are adapted to a life underwater. They spread by
horizontal underground stems called rhizomes. Once a
seagrass seed sprouts, the plant can spread by rhizomes
and colonize an entire sandflat. Seagrass supports interesting animal communities and aids, as do the mangroves,
in providing energy that helps sustain the coral reef. The
food webs of mangroves, seagrass, and coral reef are all
interlinked.

Types of Seagrass

The most common seagrass of the tropics is turtle grass
(*Thalassia testudinum*), which has long, flattened green
blades. Another species, less common than turtle grass, is
manatee grass (*Syringodium filiforme*), which has rounded
leaves resembling long thin cylinders. Both turtle grass
and manatee grass can occur singly or in mixed clumps.
Both the names *turtle grass* and *manatee grass* are well chosen. Indeed, sea turtles and manatees (*Trichelchus mana-
tus*) frequent seagrass flats and feed off the grasses.
Among the grasses are clumps of various green algae,
many of which are hardened by calcium deposits. *Penicillus* resembles a shaving brush protruding from the sand.
Udotea is bright green, suggesting a small funnel an-

chored in the sand. *Halimeda* resembles thick clumps of green disks.

A Swim Among Mangroves and Seagrass

After our strenuous walk through the mangrove caye, it is time for a refreshing swim. The Caribbean is irresistible as we don face mask, snorkel, and fins and plunge into the 80°F water. The shallow, sandy bottom is a bit disturbed by the turbulence we create, and some coral sand momentarily clouds the water. Soon we swim into clarity and begin to inspect the inhabitants of the submerged mangrove roots.

Attached to the roots are numerous oysters, sponges, barnacles, and tunicates, all feeding by filtering tiny plankton from the sea. Hardly any bare area can be found on a submerged root. Space is at a premium for sessile marine invertebrates. Among the attached animals crawl the thin brittle stars that randomly wave their five arms if picked off their chosen root. Tiny colorful coral reef fish swim in and out among the stilted roots. Black striped sargeant majors (*Abudefduf saxatilis*) poke at the roots, as a school of juvenile French grunt (*Haemulon flavolineatum*) accompanied by a few juvenile schoolmasters (*Lutjanus apodus*) swim hastily past. Their speed may have something to do with the small great barracuda (*Sphyraena barracuda*) lurking in the shade of the roots. The mangroves provide food and shelter for many of the juvenile fish that, as adults, will populated the coral reef.

As we swim away from the mangroves and over turtle grass, we feel a rhythm of gentle rolling, a mild swell waving blades of marine grass back and forth. There are shells of large queen conch (*Strombus gigas*) scattered about. Many of the shells are uninhabited by conchs but are instead the domains of tiny damselfish. These brilliant violet and yellow fish are so loyal to their conch shell that it is possible to catch them merely by lifting the shell from the water. They swim into it! Large reddish orange sea stars (*Oreaster reticulatus*) are scattered here and there, as are the foot-long, orange-brown sea cucumbers (*Isostichopus badionotus*). These are two of the most conspicuous echinoderms of the seagrass beds. Schools of parrotfish of several species graze like herds of aquatic cows over the turtle grass. They make small puffs as they spit out sand that they ingested as they munched the grass. Listen closely and you'll hear a staticlike sound as the many fish mouths grind coral, converting it into sand. More schools of grunt swim by as well as a large school of silvery bonefish (*Albula vulpes*). As we continue over the turtle grass, we note scattered coral heads. Over there is boulder coral (*Montastrea annularis*). To our left is some elkhorn coral (*Acropora palmata*). Among the corals we see more of the various coral reef fish species. A small school of reef squid (*Sepioteuthis sepioidea*) dart by, disturbing some needlefish (*Strongylura notata*) at the water surface. As we swim back to the boat, we spot a small hawksbill turtle (*Eretmochelys imbricata*) swimming over beds of turtle grass. How appropriate.

THE FOOD WEB OF COASTAL TROPICAL ECOSYSTEMS

Both mangroves and seagrasses accomplish high productivity. Red mangrove can produce 8 grams of dry organic matter per square meter per day, a rate of carbon fixation considerably higher than most other marine or terrestrial communities (Golley et al. 1962). Leaf and twig fall in a red mangrove forest was estimated to be over 3 tons per acre per year.

This productivity is not wasted. When a mangrove leaf or turtle grass blade drops off, it is almost immediately colonized by a wide range of bacteria, fungi, and protozoans. The colonization by microbes results in increasing the protein concentration of the fallen leaf (because the bodies of the microbes on it are rich in protein and because microbes consume carbohydrates, thus reducing the ratio of carbohydrate to protein), and the leaf will then be grazed by shrimp, worms, crabs, and various fish, including juvenile grunts and snappers. As the leaf is chewed, the microbes and part of the leaf are digested, but the remainder of the leaf either is spit out or passes through the alimentary canal. In either case it is recolonized by microbes, and the process begins again. In this manner the vast amount of energy and minerals trapped by the mangroves and seagrasses is slowly and steadily released to the animal inhabitants of the community. Mangroves and seagrass export both energy and nutrients to neighboring estuaries and coral reefs by providing a constant source of food for invertebrates and juvenile fish.

CORAL REEFS

In warm tropical seas around the globe, coral reef surrounds islands and parallels coastline. All reef-building coral species are confined to tropical waters where water temperature does not drop below 20°C. Coral reefs rival rain forests in productivity. Like rain forest, coral reef abounds with coevolutionary associations among species, including the very animals that give the reef its name. Fortunately for the snorkeler or scuba diver, most coral reef inhabitants, in particular the colorful fishes, are relatively easy to observe. Kaplan (1982) provides an introduction to the ecology of Caribbean reefs and identification of their inhabitants.

Coral is an animal, a member of the phylum Coelenterata, to which belong jellyfish, sea anemones, and hydras. Each coral animal is a tiny polyp housed in a calcium carbonate test of its own creation. The polyp is sessile and feeds by capturing tiny food particles (the reef plankton) in a ring of minute tentacles armed with poison cells. The cumulative efforts of many millions of coral animals build the reef. Coral animals are usually nocturnal, and the tiny coral heads embedded in the reef matrix are therefore often not evident to the snorkeler during the day.

Photosynthesis is done by plants. Coral reef accomplishes nearly as much productivity per unit area as rain forest (Whittaker 1975). However, if coral reef is such a productive ecosystem, where are the plants? What are the coral reef equivalents of the giant rain forest trees? Animals abound, fish seemingly everywhere, often in large schools. Plants, by contrast, seem conspicuous by their absence. Though it may seem paradoxical, plants on a coral reef are at the opposite end of the size spectrum from rain forest trees. The plants responsible for photosynthesis on a reef are mostly microscopic, one-celled algae that live among the matrix of reef material secreted by the coral polyps. These tiny organisms, most of which are dinoflagellates called zooxanthellae, live in intimate association with coral animals. Supplied with shelter and waste products from coral, zooxanthellae photosynthesize in the bright light that penetrates the clear waters over the reef. Coral animals use both the oxygen produced by the zooxanthellae as well as some of their products of photosynthesis. Zooxanthellae and coral represent an evolutionary mutualism: both groups of organisms are interdependent. Their interaction ultimately supports the bulk of the organisms that inhabit the reef.

There are approximately 60 coral species in the Caribbean, compared with over 700 species in the Indo-Pacific. This pattern of species richness is paralleled by fish. There are approximately 500 species in the Bahamas but just over 2000 species in the Philippines and 1500 species on the Great Barrier Reef off eastern Australia. Although the richnesses of fish communities in the tropical Atlantic and Pacific differ, the pattern of species diversity is about the same for both areas (Gladfelter et al. 1980). This means that any small area of reef in the Caribbean will contain about the same number of fish species as an equal area, say, on the Great Barrier Reef. Fish diversity correlates with reef surface area, surface complexity, and height. However, the "pool" of species (no pun intended) is much greater in the Pacific, so the between-habitat diversity is greater there than in the tropical Atlantic. It is not clear why species richness is so much lower in the Caribbean, but one possibility is that the Atlantic is a much younger ocean than the Pacific, and, thus, there has been less time and less area for speciation to occur. Caribbean coral reefs are present from the Florida Keys through the Bahamas, Greater and Lesser Antilles, and along the Yucatán Peninsula.

Reefs grow as generation after generation of individual coral polyps add to the matrix. The living part of the reef rests atop the efforts of thousands of past generations. Three types of reef occur, and one type may, in time, develop into another. Fringing reefs surround most tropical islands. Atolls are also rings of coral, often incomplete, but no island is present. However, atolls are the remains of fringing reefs that surrounded an island that is now submerged. Only the ring of coral remains visible from the surface. From observations made during the *Beagle* voyage, Charles Darwin recognized that fringing reefs may become atolls (Darwin 1842). As an island subsides, the coral animals continue to build the reef. The island sinks, but the reef doesn't. Another type of reef, barrier reef, runs parallel to a coastline. The longest barrier reef in the western hemisphere is 200 kilometers, running the length of the Yucatán Peninsula.

Coral species are adapted to different disturbance regimes. Some tolerate wave action better than others. Some do best in shallow waters, some in deeper waters. Because of wave action, differing tolerances and interactions among coral species, and relationships between coral and zooxanthellae (see below), reefs tend to be zoned.

The most windward part of a reef is often termed the *high energy zone*. It is here that the reef is most exposed to wave action. The reef front or fore reef is the area from the windward edge of the reef to the lower limits of coral growth, normally in excess of 70 meters deep. A swimmer moving toward the windward edge swims over shallow coral formations until coming to a point where the reef drops off abruptly, and only deep blue can be seen looking down through the facemask. If the swimmer parallels the beach, swimming over the reef, a structural pattern called *spur and groove* formation can be observed. Shallow walls of coral are interspersed with deeper canyons where coral sand and scattered brain corals and other species cover the bottom. The spurs are the dense coral walls that have built up fringing the mainland or island. The grooves are canyons created by erosion from strong wave action. Many coral species are common in this zone, including elkhorn and staghorns (*Acropora*), boulder coral (*Montastrea*), finger coral (*Porites*), brain coral (*Diploria*), starlet coral (*Siderastrea*), sheet coral (*Agaricia*), and stinging coral (*Millepora*; stinging coral is not a true coral—it is a hydrozoan, related to the Portuguese man-o'-war).

The next zone is the windward reef flat. This is a shallow area on the reef crest, often dominated by stinging coral, various soft corals (*Palythoa* and *Zoanthus*), and encrusting red algae. This zone also typically contains scattered elkhorn and staghorn corals, boulder coral, and other. Coral sand covers the bottom, and scattered green algae (i.e., *Halimeda*, *Penicillus*, *Acetabularia*) are present.

Further leeward is the lagoon zone. Lagoons are protected areas, usually shallow (10–15 meters), that support populations of turtle grass, scattered coral heads, and green algae. Mats of blue-green algae (cyanobacteria) may also occur. Sea stars, sea cucumbers, and conchs are common.

Coral species tend to exhibit vertical zonation. The deepest corals are encrusting species that occur at depths of up to 60 meters. Hemispherical corals, such as brain corals, occur at depths of 5–35 meters, while branching corals, such as staghorn and elkhorn coral, are shallow, from surface to 10 meters. James W. Porter (1976) found that shallow water corals are highly branched and have a high surface area/volume. Further, their average polyp diameter is small compared with deeper water species. The geometric growth pattern exhibited by shallow corals is that characteristic of multilayered trees! Elkhorn and staghorn corals have high concentrations of zooxanthellae and are functioning metabolically as autotrophs. They function as plants and their growth forms reflect this.

Hemispherical corals have large polyps, less zooxanthellae, and a smaller surface area/volume ratio. They are metabolically more heterotrophic than autotrophic. Deep water corals have large polyps and are total heterotrophs.

The diversity of coral species in any given zone is strongly influenced by disturbance history (Woodley et al. 1981, see page 155) and interactions among the coral species (Porter 1972). Many examples of competition are known for corals. For example, stinging corals (*Millepora* spp.), often redirect their growth in the direction of gorgonian (sea fan) corals (*Gorgonia* spp.). The stinging coral contacts, abrades, encircles, and encrusts the "target" gorgonian (Wahle 1980).

A comparison between rain forest and coral reef reveals some striking similarities. Both ecosystems are highly productive, virtually the most productive of any natural ecosystems. Net primary productivities can approach 3000 grams dry matter annually, sometimes higher. One difference, however, is that primary producers, the plants, are far more diverse and have vastly more biomass in rain forest. One-celled algae perform the same role, with the same success, as giant rain forest trees. There is a high species richness of animals in both coral reef and rain forest, including a high predator diversity. One difference, however, is that animals are generally easier to observe on reefs, though many nocturnal species occur. Territoriality is common among animals in both ecosystem types. Sexual selection is also evident. Fishes, such as wrasses and parrot fish, are highly sexually dimorphic and polygynous. Just as mixed flocks of birds forage in rain forests, mixed species schools of fish exploit coral reefs. There are numerous examples of coevolution in both ecosystems. Some coral reef fish are called "cleaners." They move into the mouths and around the skins of larger fish and remove attached parasites. There are also species that mimic cleaner fish but bite a chunk out of the host fish instead (Limbaugh 1961)! Both coral reefs and rain forests contain many cryptic species and also some species exhibiting warning coloration. Toxic compounds and allelochemics are evident among species from both ecosystems. Rain forest species show both a high within- and high between-habitat species diversity. Coral reefs in the Caribbean lack a high between-habitat diversity, possibly because of historical factors. Diversity may be partly a result of nonequilibrium in both ecosystems. Long periods without disturbance may result in lowered diversities in both systems as competitive exclusion occurs.

CHAPTER 10

Epilogue

> In many ways, Brazil today is reminiscent of the United States of generations ago. Vast areas of the country are unoccupied. Were the potentials of the country realized, it could probably sustain more people at a higher standard of living than could the United States.
>
> —Philip H. Abelson, 1975

> If we ever allow tropical forests to disappear from Earth, we shall have to tell ourselves not only that we have lost something of value. We shall have to admit that in certain significant senses, we have not gained much since we left our caves. For all that we are now firmly established on our hind legs, our response to tropical forests serves as a measure of how far we are ready to stand tall as citizens of the Earth, as joint members of our planetary home.
>
> —Norman Myers, 1984

THESE divergent views reflect a current debate about the fate of the tropics. Some wish to rapidly develop the tropics, cutting forests to use the land for a variety of human needs. Others believe that the tropics are in imminent danger of being misused, their resources depleted. Still others opine that rain forests comprise a unique biological resource, a refuge for biological diversity that should be preserved for its own sake. The popular press has been peppered with editorials and stories about the cutting of global tropical forests. Periodicals from *National Geographic* to the *New Yorker* have published feature stories on the ecology of tropical forests and the struggle over their fate.

Though opinions vary, one fact is clear. Tropical forest ecosystems are being converted into other systems at unprecedented rates. The world's tropical forests are being felled for lumber, roads, pastures and cropland, and tree plantations. Rates of extinction may be higher today than for most of the planet's history, largely because of what is occurring in the tropics. The issues usually represent a clash between short-term economic gain versus long-term ecological loss.

With the possible exception of the Amazon Basin, still a vast region of undeveloped tropical forest, the American tropics are rapidly changing. Someone visiting the Neotropics for the first time is apt to be surprised by the paucity of lowland forest. Cattle pastures are not hard to find. Howler monkeys can be.

CASE STUDY: RESETTLEMENT OF PEOPLE IN THE BRAZILIAN AMAZON

Daniel Janzen (1972, 1973) pointed out how different the tropics are from the temperate zone and how failure to consider these differences often has resulted in failed agriculture and habitat deterioration. No example better illustrates this point than the attempt of the Brazilian government to relocate large numbers of people to interior Amazonia.

The Amazon forest seems almost sirenlike to those with visions of exploitation. The forest, so lush, so expansive, seems to call out, offering vast potential for development. The Brazilian government heard the call and attempted to develop Amazonia, initially by making it accessible through construction of a transamazon highway, followed by the resettlement of 1 million families from the impoverished northeastern region of the country to this newly opened, promised land. Nigel J. H. Smith (1981) reported that after ten years of effort, only 8,000 families, far below the original goal of 1 million, were actually resettled. The effort has been fraught with difficulties ranging from unforeseen ecological effects to spread of diseases by the colonists themselves. Smith's study, which I outline below, documents the difficulties inherent in using conventional methods of landscape alteration in Amazonia.

The Transamazon Highway was opened in 1975. It is a two-lane unpaved network of roads making the western Brazilian Amazon accessible from Belém and Brasilia. The road was opened with the hopes not only of resettling many families but also of exploiting the region for minerals and timber and to demonstrate to would-be foreign developers that Brazil was "in charge" of its own real estate.

The highway cuts through interior forest, well off of flood plain. Soils are nutrient-poor, with high concentrations of toxic aluminum. A few fertile soils occur near the highway, but overall, only 3% of the soils in the region are fertile. As a result, to farm effectively, fertilizer is required and fertilizer costs money. After cutting and burning, high rainfall in the region subjects the exposed land to rapid erosion. Not only do farmers' fields erode, but so does the Transamazon Highway. Side roads may be cut off from main roads during the rainy season, and the main roads may be damaged, cutting off suppliers from the cities.

The intent of the Brazilian government was for the newly arrived farmers to plant upland rice as their basic crop. This was an unfortunate decision on the planners' parts since upland rice requires fertilizer, herbicide, and insecticide treatment to remain healthy. The impoverished farmers were not in any position to buy such expensive agricultural aids. To make matters worse, poor strains of rice were selected. The commonest strain, IAC 101, was developed barely within a tropical climate and was subject to stem breakage and flattening during heavy rains in the wet Amazon. Fungi invaded, rotting the flattened plants, resulting in dramatically reduced yields. The average farmer, though he planted eight hectares annually, only grossed $1900 for his crop. Rats, ground doves, fungi, and erosion claimed a large share of each crop.

Manioc (page 103) would have been a better choice. This species, *Manihot esculenta*, is adapted to the wet climate and poor soils. An estimated annual yield of 20 tons per hectare could permit a family that can harvest 3 hectares to earn $3000 from the sale of flour made from manioc. The harvestable part of manioc, the tuberous root, can be collected anywhere from six months to three years following planting. Farmers can cooperate and aid each other's harvest since harvest time is not critical, as it is with rice. Roots can remain unharvested during the worst of the rainy season when roads are flooded. Manioc flour is easily stored, and the food itself yields more calories than rice.

The Brazilian government was slow to embrace manioc as a staple crop. Instead it attempted to introduce new crops, not only rice, but also pepper, banana, and cacao. It also promoted the growth of cattle ranching along the highway. Though the region did not initially harbor agricultural pests, they soon began to appear. The highway, lined with newly planted crops, served as a corridor for the invasion of pest species into the interior Amazon. Pepper plantations were injured by a fungus that appeared in 1975 and was resistant to fungicides. Fungi also attacked the numerous banana plantations along the highway.

People brought their own diseases. In 1973, reports indicated that regional public hospitals had treated 1285 cases of malaria. The rain forest itself did not initially harbor the malarial parasite, a minute protozoan called *Plasmodium* (page 63). After settlement, three species of malaria were reported from the region, and all three were brought from the north by the people themselves. The mosquito vector of the parasite, *Anopheles darlingi*, thrived in the water culverts bordering the highway, and the disease became very common. Unfortunately, malaria has two annual peaks, one of which occurs when the farmers are sowing their crops, and the other approximately at harvesttime. Such timing couldn't be worse from the standpoint of productivity. In addition to malaria, gastrointestinal diseases and parasite infections derived from poor sanitation also plague the people. Diarrheal diseases are the most serious causes of infant mortality in the region.

The colonization of the interior Amazon cannot be called a success at this time. Only eight sawmills have been built, and the valuable commercial tree species are already largely removed. No great mineral deposits have been found. The overpopulated northeastern region of Brazil has grown by 6 million since the Transamazon Highway was built, and relocation of the people has accounted for less than 1% of the northeastern population growth. The cost to Brazil has been approximately $500 million, a very steep price for a project with so few benefits. As Smith wrote in his paper, "Along the Transamazon, many settlers soon fell victim to a biased and inefficient credit system, a poor selection of crops, infertile soils, and isolation from large markets."

In an ambitious attempt, Philip M. Fearnside (1986) developed a complex computer model to predict sustainable human carrying capacity of Brazilian forests under different possible scenarios. His rather pessimistic conclusion was that potential carrying capacity for people under both simple agriculture as well as under larger enterprises, such as cattle ranching, is disappointingly low.

THE PLIGHT OF THE GLOBAL TROPICS

Rather than being clothed in dense jungle or tall pristine rain forest, most of the tropics, not only in the Americas but in Africa and Asia as well, consists of pastures, agricultural fields, timber plantations, and an ever-increasing network of roads (Myers 1980, 1984). In its report, *Research Priorities in Tropical Biology* (Committee on Research Priorities in Tropical Biology 1980), the National Research Council (NRC) estimated that approximately 16 million square kilometers of tropical moist forest exist potentially on earth. However, by 1975 the actual area covered by moist forest was only 9.35 million square kilometers, a reduction of 41.5% from the potential. Approximately 200,000 square kilometers of tropical moist forest are being converted annually, amounting to the cutting of 20–40 hectares per minute! The NRC estimated that an area the size of Great Britain is cut annually and an area the size of Delaware is felled weekly. Already, two-thirds of the forests in India, Sri Lanka, and Burma have been removed, and the remainder of the tropical world is rapidly approaching that figure. The NRC estimates that by the year 2000 only in remote parts of the Amazon Basin and the African Congo will there exist uncut rain forest. Examined on a regional basis, by 1975, 51.9% of African moist forests had been converted into some other kind of ecosystem compared with 43.7% in Asia and 37.0% in the Neotropics. However, current conversion rates are highest in Latin America with an estimated 50,000–100,000 square kilometers being cut annually compared with 20,000 per year in Africa and 50,000 per year in Asia.

International trade has stimulated clearance of tropical forests. In one example, termed the "hamburger connection," rising beef costs in the United States stimulated importation of foreign beef. Beef imports were increased by 7.6% in 1978 and an additional 5% in early 1979. Most imported beef is range-fed from Latin America and is considered satisfactory only for fast-food chain restaurants or as supermarket ground beef. These additional beef imports, mostly from the Neotropics, have been estimated to reduce by a nickel the cost of a hamburger at a fast-food restaurant. To quote from the NRC report:

> Therein lies the connection between difficulties of the U.S. domestic economy and the demise of TMF [tropical moist forest] in Central America and Amazonia. The price of a U.S. hamburger does not reflect the total costs, and especially the environmental costs, of its production in tropical Latin America; and the American

consumer, seeking best quality hamburger at the least price, is not aware of the ramifications in forest zones thousands of kilometers away.

It is easy to understand why cattle pastures are so rapidly replacing forest tracts throughout the Neotropics.

Human population growth is higher in tropical areas than elsewhere on earth, exacerbating the destruction of tropical forests. It is estimated that 90% of the world population growth during the next twenty years will occur in the tropics. By the turn of the century, South America will increase from 193 million persons to 332 million. As prime land is converted to agriculture and pasture, large numbers of peasant farmers overtax what little land they may own, or they are forced to farm less optimal areas. Of the 800 million people estimated by the World Bank to be living in conditions of absolute poverty, the vast majority reside in the tropics. Absolute poverty is defined as a "condition of life so characterized by malnutrition, illiteracy, disease, squalid surroundings, high infant mortality, low life expectancy, as to be beneath any reasonable definition of human decency." The annual global growth rate of the human population is 1.7%, but in west Africa the figure is 3.0%, in southeast Asia it is 2.2%, and in Latin America it is 2.8%. Annual growth rate of the U.S. population is 0.6%; of the USSR, 0.8%; of Japan, 0.9%; and West Germany, 0.2%.

There are many areas in which tropical research could provide important information for human welfare. Of the approximately 4.5 million species of animals and plants estimated to exist, about two-thirds occur only in the tropics. Of these nearly 3 million tropical species, only about 500,000 have been described and named. Some taxonomic groups like birds, for instance, are relatively well described, although new species continue to be discovered even among these groups. Many taxonomic groups are poorly known. Many tropical organisms have limited ranges and are, thus, highly prone to extinction should their restricted habitats be lost. An example of the sort of problem potentially generated by species loss is the case of teosinte, a perennial wild maize (corn) found only in an area of less than two hectares in Jalisco, Mexico. According to the National Research Council, this maize species is virus-resistant, with obvious potential for agriculture. With the advancement of genetic engineering, the ability to move beneficial genes from one organism to the next is looming in the near future. Tropical species represent potential reservoirs of genetic variability, combinations of genetic material that could have many beneficial applications. Because of the rapid removal of mature forest ecosystems, and the resultant high rates of extinction that will ensue, many species may disappear before we even know they existed.

Although tropical moist forests are the most species-rich terrestrial ecosystems on the planet, it is not at all clear how this extremely high diversity is generated and maintained (chapter 4). Ecologists are attempting to ascertain just how fragile tropical forests are. If substantially disturbed, how rapidly do they recover? At what level of disturbance do they not recover? Tropical forests do not function in ways identical to forests in the temperate zone. Many tropical soils are delicate, prone to rapidly lose fertility if their forest covering is removed. Most nutrients are in biomass, not the soil bank. Rates of reycling are extraordinarily rapid but can be easily disrupted (chapter 3).

Current research suggests that improved management and use of fertilizers, especially phosphate, can substantially enhance productivity. In eastern Peru, Abelson (1983) noted that fertilization permitted a three crop per year agriculture, without loss of soil fertility. Arturo Gómez-Pompa and his colleagues (1982) have experimented with hydraulic agriculture, similar to that utilized by ancient Mayans (see chapter 3). This methodology involves the construction of *chinampas*, or canals that raise fields for agriculture. Through manual labor alone, crops such as manioc, radishes, lettuce, squash, rice, corn, watermelon, chiles, turnips, tomatoes, cucumbers, coriander, parsley, swiss chard, and beans were successfully raised. The surrounding canals could be utilized to raise fish and turtles, exactly as they were during classic Mayan time. Hydraulic agriculture reduces erosion and relies on manual labor, a commodity in great supply in overpopulated lands.

POTENTIAL GLOBAL EFFECTS OF DEFORESTATION

An insidious problem may ensue as a result of cutting and burning tropical forests. Burning releases carbon dioxide into the atmosphere. George Woodwell (1978) points out that carbon dioxide levels have increased from 290 ppm (parts per million) prior to the Industrial Revolution to 320 ppm by 1970. This increase was due mostly to the burning of fossil fuels. However, from 1970 to 1980, the carbon dioxide concentration in the atmosphere increased yet another 10 ppm, to 330 ppm. Some studies suggest that perhaps half of the 10 ppm increase over the decade is due to carbon dioxide liberated from burning of felled tropical forests. The remainder was from burning of fossil fuels in industry, home, and transportation. Increasing amounts of atmospheric carbon dioxide will enhance the "greenhouse effect," the warming of the earth. Carbon dioxide traps heat being radiated from the earth's surface, making global climate warmer and, some believe, more variable. Variations in climate may result. Some areas may have hotter summers and colder winters, some desert areas may shrink or increase. Tropical forest removal not only adds carbon dioxide to the atmosphere, it also keeps it there.

One difficulty is the accuracy with which the global carbon cycle can be determined. Huge amounts of carbon are involved in a dynamic system involving marine and terrestrial ecosystems plus burning of fossil fuels. Recent research suggests that total carbon released from conversion of tropical forests is lower than previously estimated and that the global carbon cycle will likely remain in balance (Detwiler and Hall 1988). Shifting cultivation is still prevalent in the tropics and results in substantial ecosystem recovery, attenuating the initial effects of forest cutting.

Nonetheless, effects of forest clearance can be subtle. Even increased cattle grazing could conceivably affect global climate. Cattle produce much methane in the course of their digestion of plant fiber. When methane is liberated by cattle burps, it, too, contributes to greenhouse effect, acting to trap heat as carbon dioxide does. Termites, which also increase when forests are felled and pastures created, contribute as well to the build up of methane, carbon dioxide, and other greenhouse effect chemicals (chapter 7, page 319). Many questions concerning the possible effects of deforestation on climate exist. The final results of global tropical deforestation may range well beyond the extinction of tropical species.

One recent study links deforestation with increased flooding in upper Amazonia. A. H. Gentry and J. Lopez-Parodi (1980) report that the height of the annual flood crest at Iquitos, Peru, has increased dramatically over a decade. The increase correlates with greatly increased deforestation in the region. The study supports the claim that regional levels of precipitation are dependent on transpiration from plants. Thus far, no significant changes have been seen in regional patterns of precipitation. It rains the same amount, but there is much less watershed to absorb the rain, thus it directly drains to the Amazon. A related problem is occurring in Panama. James Karr told me that a high rate of sedimentation is occurring in the Panama Canal. Deforestation is contributing to erosion of soil, which is washing into and filling the canal.

Thus far, probably due to its utter immensity, the Amazon Basin remains in equilibrium in terms of its water and mineral cycles. Eneas Salati and Peter B. Vose (1984) report that although deforestation has been "very active" over the last decade, the Amazon forest still efficiently recycles at least 50% of the precipitation that falls on it. Not only that, but minerals are being held by the system, rather than eroded away. Salati and Vose did warn that flooding, especially in the lower Amazon, is likely to increase as deforestation continues. Ultimately, precipitation levels could drop because there are too few trees to recycle the water. Should this occur, climate and agricultural productivity could be adversely affected.

A HOPEFUL NOTE

Though the overall picture for the future of the Neotropics may seem bleak, some are less pessimistic. Michael A. Mares (1986) points out that South America now has 218 parks and reserves containing 488,906 square kilometers of habitat, a figure that represents 2.7% of the continental land area. He argues that most South American countries have tried, within their limited economic means, to create protected national parks, and that a coordinated and massive effort of governments and specialists in conservation biology would secure future protection. Tropical conservation faces serious problems: lack of data, lack of trained conservation experts, lack of money, lack of long-term conservation policies, weak economies, short-term economic strategies, and an air of panic over the situation in general. Nonetheless, Mares believes there is still time to resolve the problems, arguing that extinction rates may be less than initially feared. There are no easy solutions to balance human needs against those of the rest of the biota. J. Marcio Ayres (1986) points out that the actual implementation of conservation-oriented policies is often slow and fraught with social problems. Ayres asks, for instance, "How can hunting be banned in Amazonia, when equivalent cattle meat would cost at least 10% of the average salary?" The price paid a hunter for a jaguar skin ($70) is equivalent to 88% of the regional minimum monthly wage.

The Nature Conservancy, using its natural heritage programs as a model, has helped create a series of CDCs, or Conservation Data Centers, throughout the Neotropics (Norris 1988). Eight CDCs are scattered in as many countries. These centers inventory species, documenting what is rare and what is common.

Creative efforts are underway to secure national parklands in the Neotropics. A major project in Costa Rica, largely spearheaded by Daniel Janzen, is directed at creating Guanacaste National Park (McLarney 1988). This park has the support of the Costa Rican government as well as the local people. Thus far, over $3 million have been raised, much of it foreign donations, to acquire the necessary land. One unique plan, developed by Conservation International (CI), exchanged a small portion of Bolivia's foreign debt (acquired by CI) for Bolivia's commitment to conserve four million acres around El Porvenir biological station, which CI helps operate (Truell 1988). Worldwide efforts to support tropical conservation are producing results. Janzen (1988) has articulated a view in which he equates the value of wild ecosystems with such societal needs as "libraries, universities, museums, symphony halls, and newspapers." Janzen argues that tropical peoples are sympathetic and supportive of conservation efforts.

A FINAL WORD

I hope this book has interested you in tropical ecology and has impressed you with the need to be aware of the potential long-term value of tropical ecosystems, especially the forests. Such awareness, if widespread, could help temper the rapid destruction of tropical forests. In a paper published in 1972, A. Gómez-Pompa, C. Vazquez-Yanes, and S. Guevara called the tropical rain forest a "nonrenewable resource." They urged immediate international efforts be taken to protect "this gigantic pool of germplasm by the establishment of biological gene pool reserves from the different tropical rain forest environments of the world." Today, their calls for action are perhaps beginning to be heeded.

The future of the American tropics?

GLOSSARY

AIR PLANT. A plant that lives on another plant but does not parasitize it. Examples are lichens, bromeliads, orchids, etc. Same as epiphyte.

APOGEOTROPIC ROOTS. Roots that grow upward from the soil on the trunk of another tree.

ARUM. A group of plants to which philodendrons belong. Florescence organized in an arrangement where a leafy petal called a spathe surrounds a central spike of flowers called a spadix.

BATESIAN MIMICRY. A situation in which a palatable animal species comes to resemble an unpalatable species, thus gaining some protection from predation.

BIOGEOCHEMICAL CYCLING. The continuous and cyclic movement of nutrient atoms such as calcium, nitrogen, phosphorus, and magnesium between the living and the nonliving components of the ecosystem.

BIOMASS. The total weight of living material in an ecosystem.

BROMELIAD. A type of epiphyte characterized by a basal cluster of spikelike leaves and an elongate flower stalk.

BUTTRESS. A tree root that extends out from the trunk as a phlangelike structure.

CANOPY. The upper layer of foliage in a forest.

CAULIFLORY. The characteristic of having inflorescences (and thus fruits) grow directly from the main trunk of a tree.

COEVOLUTION. The evolutionary interaction of two or more species acting as selection pressures on each other.

COMMUNITY. The total assemblage of plants, animals, and microbial organisms that interact in a given ecosystem.

CLOUD FOREST. A mountain forest that exists in perpetual mist, characterized by stunted trees with an abundance of epiphytic growth.

CRYPTIC COLORATION. Camouflaged appearance rendering the animal less visible.

DECIDUOUS. The characteristic of dropping leaves during periods of stress due to dryness.

DEFENSE COMPOUND. A type of chemical synthesized by plants that confers some protection against herbivores. Common examples are terpenoids and phenolics.

DRIP TIP. The sharply pointed tip of a typical tropical leaf, named for its tendency to drip rain water.

ECOSYSTEM. The total interacting living (biotic) and nonliving (abiotic) components of a given area.

EDAPHIC. Pertaining to the soil.

EMERGENT. A very tall tree that exceeds the average canopy tree in height.

EPIPHYTE. See Air plant.

EXTRA-FLORAL NECTARIES. Nectar-rich bodies present on many tropical plants that are fed upon by ants and wasps but that are not flowers.

FRUGIVORE. An animal that eats primarily fruit.

GALLERY FOREST. A generally lush forest that grows along a riverbank and floodplain.

GAP. An opening in the canopy where light intensity is high.

HERBIVORE. An animal that eats only plant material, such as a caterpillar.

HUMUS. The complex organic material resulting from the decomposition of forest leaf and branch litter.

JUNGLE. A tangled, dense successional ecosystem consisting of many fast-growing, light-loving species.

LATERITE. A kind of tropical soil high in aluminum and iron compounds, often reddish in color.

LATERIZATION. The process whereby leaching caused by heavy rains plus high temperatures converts lateritic soils into hardened bricklike material.

LEACHING. The removal of nutrient atoms from soils brought about by the effect of rainwater washing through the soils, interacting with the clay component of the soil.

LEK. An area in which several, and sometimes several dozen, males court passing females.

LIANA. A type of woody vine that begins as a small shrub and entwines upward, entangling throughout the canopy.

LIFE ZONE. A recognizable band of vegetation along a mountain slope.

MACHETE. A long-bladed knife used to cut tropical vegetation.

MANGROVE. A group of tropical woody plants that are highly tolerant of immersion in salt water. They comprise a major ecosystem type along tropical coasts.

MIMICRY. A situation in which an organism, either through appearance, behavior, or both, comes to closely resemble another unrelated species that shares the same ecosystem. See Batesian mimicry, Mullerian mimicry, and Mimicry complex.

MIMICRY COMPLEX. A situation in which a group of several butterfly species, including some from different taxonomic families, converge in appearance.

MIXED FLOCK. A foraging flock of birds comprised of several, and sometimes many, species.

MOIST FOREST. A seasonal tropical forest receiving not less than 100 millimeters (approximately 4 inches) of rainfall in any month for two out of three years, frost-free, with an average temperature of 24°C or more. See Rain forest.

MONTANE. Mountainous.

MÜLLERAIN MIMICRY. A situation in which two or more unpalatable species converge in appearance.

MUTUALISM. A situation in which two or more species become evolutionarily interdependent in such a way that each benefits the other(s).

MYCORRHIZA. A group of fungal species that are mutualists with trees. Fungi, which grow into the roots, aid in taking up minerals from the soil. Trees supply fungi with carbohydrates from photosynthesis.

NATURAL SELECTION. The mechanism of evolutionary change first formally described by Charles Darwin and Alfred Russel Wallace. Argues that genetic characteristics best suited to a particular environment will be disproportionately passed to offspring because these characteristics confer greater survival value, and, thus, their possessors reproduce in greater numbers.

NEOTROPICS. The term for the American tropics.

PAMPAS. Temperate grasslands in the central and southern parts of South America.

PARAMO. Mountain shrublands found at higher elevations throughout the Andes.

PHOTOSYNTHESIS. The complex biochemical process by which green plants capture a small amount of the sun's light energy and incorporate it, with water and carbon dioxide, into energy-rich sugar compounds.

POPULATION. A group of individuals all of which are the same species.

PREHENSILE. A type of tail found on certain Neotropical monkeys and opossums that functions as a fifth limb. The tail is capable of being curled around a branch, holding the animal securely.

PRODUCTIVITY. The total amount of photosynthesis that occurs in a given ecosystem.

PROP ROOT. A root that leaves the trunk or a branch well above ground and helps anchor the tree. Same as stilt root.

PUNA. A high elevation grassland found throughout the Andes.

RAIN FOREST. A very wet, essentially nonseasonal forest. See Moist forest.

REFUGIA. Shrunken areas of rain forest that were scattered in Central and South America during the Ice Ages. So named because rain forest species found "refuge" in these rain forests, which were otherwise surrounded by savanna.

SAVANNA. An ecosystem that is primarily grassland but with scattered trees and shrubs.

SELECTION PRESSURE. A characteristic of the environment of an organism, either abiotic or biotic, that influences the probable survival of the organism.

SEXUAL SELECTION. The process first described by Charles Darwin in which females mate preferentially with the most "attractive" males and/or in which males compete among themselves for females. The result of female choice and male/male competition is to select for colorful and/or large males.

SPECIES RICHNESS. The number of species (of a given taxon) within a given ecosystem.

STRATIFICATION. The organization of trees in a forest in different horizontal layers such as canopy, subcanopy, shrub, and herb layers.

SUCCESSION. An ecological process in which groups of fast-growing species colonize a disturbed area, eventually to be replaced by groups of slower-growing species that are good competitors and that occupy the site indefinitely.

SURFACE ROOTS. Roots that radiate out on, rather than in, the soil.

TAPROOT. A thick, deep central root.

TAXON. A group of organisms that are members of the same evolutionary group. For example, birds represent a taxon, as do mammals, insects, and flowering plants.

TAXONOMY. The science of the classification of organisms.

TERRA FIRME. An area of rain forest not subject to flooding.

TRANSPIRATION. A metabolic process whereby plants take up water and minerals from the soil, evaporating the water via the leaves.

URTICATING. Hairs that contain skin irritants.

VARZEA. A term refering to riverine forests along the Amazon and its tributaries.

WARNING COLORATION. Obviousness of appearance associated with inpalatability or toxicity. Also called aposematic coloration.

REFERENCES

Abelson, P. H. 1975. Energy alternatives for Brazil. *Science* 189: 417.

———. 1983. Rain forests of Amazonia. *Science* 221, editorial.

Abrahamson, D. 1985. Tamarins in the Amazon. *Science* 85(7): 58–63.

de Andrade, J. C., and J.P.P. Carauta. 1982. The *Cecropia-Azteca* association: A case of mutualism?, *Biotropica* 14: 15.

Armesto, J. J., J. D. Mitchell, and C. Villagran. 1986. A comparison of spatial patterns of trees in some tropical and temperate forests. *Biotropica* 18: 1–11.

Askins, R. A. 1983. Foraging ecology of temperate-zone and tropical woodpeckers. *Ecology* 64: 945–56.

Augspruger, C. K. 1982. A cue for synchronous flowering. In *The ecology of a tropical forest*, edited by E. G. Leigh, Jr., A. S. Rand, and D. M. Windsor. Washington, DC: Smithsonian Institution Press.

Austin, O. L. 1961. *Birds of the world*. New York: Golden Press.

———. 1985. *Families of birds*. Rev. ed. New York: Golden Press.

Ayensu, E. S., ed. 1980. *The life and mysteries of the jungle*. New York: Crescent Books.

Ayers, J. M. 1986. Some aspects of social problems facing conservation in Brazil. *Trends in Ecol. Evol.* 1(2): 48–49.

Baker, H. G. 1983. *Ceiba pentandra* (Ceyba, ceiba, kapok tree). In *Costa Rican natural history*, edited by D. H. Janzen. Chicago: Univ. of Chicago Press.

Bates, H. W. 1862. Contributions of an insect fauna of the Amazon Valley. *Trans. Linn. Soc. London* 23: 495–566.

———. 1892. *The naturalist on the river Amazons*. London: John Murray.

Bates, M. 1964. *The land and wildlife of South America*. New York: Time Inc.

Bazzaz, F. A., and S.T.A. Pickett. 1980. Physiological ecology of tropical succession: A comparative review. *Ann. Rev. Ecol. Syst.* 11: 287–310.

Beard, J. B. 1944. Climax vegetation in tropical America. *Ecology* 25: 127–58.

Beard, J. S. 1953. The savanna vegetation of northern tropical America. *Ecol. Mono.* 23: 149–215.

Beehler, B. M., and M. S. Foster. 1988. Hotshots, hotspots, and female preference in the organization of lek mating sytems. *Amer. Nat.* 131: 203–19.

Benson, W. W. 1972. Natural selection for Müllerian mimicry in *Heliconius erato* in Costa Rica. *Science* 176: 936–39.

Benson, W. W., K. S. Brown, and L. E. Gilbert. 1976. Coevolution of plants and herbivores: Passion flower butterflies. *Evolution* 29: 659–80.

Bentley, B. L. 1976. Plants bearing extrafloral nectaries and the associated ant community: Interhabitat differences in the reduction of herbivore damage. *Ecology* 54: 815–20.

— ——. 1977. Extrafloral nectaries and protection by pugnacious bodyguards. *Ann. Rev. Ecol. Syst.* 8: 407–27.

Bishop, J. A., and L. M. Cook. 1975. Moths, melanism and clean air. *Sci. Amer.* 232: 90–99.

Bleiler, J. A., G. A. Rosenthal, and D. H. Janzen. 1988. Biochemical ecology of canavanine-eating seed predators. *Ecology* 69: 427–33.

Blydenstein, J. 1967. Tropical savanna vegetation of the llanos of Colombia. *Ecology* 48: 1–15.

Boucher, D. H. 1983. Coffee (Cafe). In *Costa Rican natural history*, edited by D. H. Janzen. Chicago: Univ. of Chicago Press.

Boulière, F. 1983. Animal species diversity in tropical forests. In *Tropical rain forest ecosystems: structure and function*, edited by F. B. Golley. Amsterdam: Elsevier Scientific.

Boulière, F., and H. Hadley. 1970. The ecology of tropical savannas. *Ann. Rev. Ecol. Syst.* 1: 125–52.

Bradbury, J. W. 1981. The evolution of leks. In *Natural selection and social behavior: Research and new theory*, edited by R. D. Alexander and D. W. Tinkle. New York: Chiron Press.

Bradbury, J. W., and R. Gibson. 1983. Leks and mate choice. In *Mate choice*, edited by P. Bateson. Cambridge: Cambridge Univ. Press.

Brandon, C. 1983. *Noctilio leporinus* (Murcielago, pescador, fishing bulldog bat). In *Costa Rican natural history*, edited by D. H. Janzen. Chicago: Univ. of Chicago Press.

Brokaw, N.V.L. 1982. Treefalls: Frequency, timing, and consequences. In *The ecology of a tropical forest*, edited by E. G. Leigh, Jr., A. S. Rand, and D. M. Windsor, 101–8. Washington, DC: Smithsonian Institution Press.

— ——. 1985. Gap-phase regeneration in a tropical forest. *Ecology* 66: 682–87.

— — —. 1987. Gap-phase regeneration of three pioneer tree species in a tropical forest. *J. of Ecology* 75: 9–19.

Brower, L. P. 1969. Ecological chemistry. *Sci. Amer.* 220: 22–29.

Brower, L. P., J.V.Z. Brower, and C. T. Collins. 1963. Experimental studies of mimicry: 7. Relative palatability of Müllerian mimicry among Neotropical butterflies of the subfamily Heliconiinae. *Zoologica* 48: 65–84.

Brower, L. P., and J.V.Z. Brower. 1964. Birds, butterflies, and plant poisons: A study in ecological chemistry. *Zoologica* 49: 137–59.

Brown, B. J., and J. L. Ewel. 1987. Herbivory in complex and simple tropical successional ecosystems. *Ecology* 68: 108–116.

Brown, L., and L. L. Rockwood. 1986. On the dilemma of horns. *Nat. Hist.* 95(7): 54–62.

Brown, S., and A. E. Lugo. 1982. The storage and production of organic matter in tropical forests and their role in the global carbon cycle. *Biotropica* 14: 161–87.

Buckley, P. A., M. S. Foster, E. S. Morton, R. S. Ridgely, and F. G. Buckley. 1985. *Neotropical ornithology.* Washington, DC: American Ornithologists' Union.

Cain, A. J., and P. M. Sheppard. 1954. Natural selection in *Cepaea. Genetics* 39: 89–116.

Canby, T. Y. 1984. El Nino's ill wind. *Nat. Geog.* 162(2): 143–83.

Capparella, A. P. 1985. Gene flow in a tropical forest bird: Effects of riverine barriers on the blue-crowned manakin (*Pipra coronata*). Abstracts of the 103d American Ornithologists' Union Meeting.

Carroll, R. L. 1988. *Vertebrate paleontology and evolution.* New York: W. H. Freeman.

Center, T. D., and C. D. Johnson. 1974. Coevolution of some seed beetles (Coleoptera: Bruchidae) and their hosts. *Ecology* 55: 1096–1103.

Coley, P. D. 1982. Rates of herbivory on different tropical trees. In *The ecology of a tropical forest*, edited by E. G. Leigh, Jr., A. S. Rand, and D. M. Windsor. Washington, DC: Smithsonian Institution Press.

———. 1983. Herbivory and defensive characteristics of tree species in a lowland tropical forest. *Ecol. Mono.* 53: 209–33.

———. 1984. Plasticity, costs, and anti-herbivore effects of tannins in a neotropical tree, *Cecropia peltata* (Moraceae). *Bull. Ecol. Soc. Amer.* 65: 229.

Coley, P. D., J. P. Bryant, and F. S. Chapin III. 1985. Resource availability and plant antiherbivore defense. *Science* 230: 895–99.

Colwell, R. K. 1973. Competition and coexistence in a simple tropical community. *Amer. Nat.* 107: 737–60.

———. 1985a. Stowaways on the hummingbird express. *Nat. Hist.* 94(7): 56–63.

———. 1985b. A bite to remember. *Nat. Hist.* 94(4): 2–8.

Committee on Research Priorities in Tropical Biology. 1980. *Research priorities in tropical biology.* Washington, DC: National Academy of Science.

Connell, J. H. 1978. Diversity in tropical rain forests and coral reefs. *Science* 199: 1302–10.

Connell, J. H., and E. Orias. 1964. The ecological regulation of species diversity. *Amer. Nat.* 98: 399–414.

Connor, E. F. 1986. The role of Pleistocene forest refugia in the evolution and biogeography of tropical biotas. *Trends in Ecol. Evol.* 1(6): 165–68.

Darwin, C. R. [1839] 1912 edition. Journal of researches into the geology and natural history of the various countries visited by H.M.S. Beagle . . . round the world under the command of Capt. Fitz Roy, R. N. New York: Appleton.

————. 1842. *On the structure and distribution of coral reefs.* London: Scott.

————. 1859. *On the origin of species by means of natural selection of favored races in the struggle for life.* London: John Murray.

————. 1862. *On the various contrivances by which British and foreign orchids are fertilised by insects, and on the good effects of intercrossing.* London: John Murray.

————. 1871. *The descent of man and selection in relation to sex.* London: John Murray.

————. [1906]. 1959. *The Voyage of the Beagle.* Reprint. London: J. M. Dent and Sons.

Davis, E. W. 1983. Preparation of the Haitian zombie poison. *Harvard Univ. Botanical Museum Leaflets* 29(2): 139–149.

Delacour, J., and D. Amadon. 1973. *Curassows and related birds.* New York: American Museum of Natural History.

Denslow, J. S. 1980. Gap partitioning among tropical rainforest trees. In *Tropical succession,* supplement to *Biotropica* 12: 47–55.

Detwiler, R. D., and C. A. S. Hall. 1988. Tropical forests and the global carbon cycle. *Science* 239: 42–47.

DeVries, P. J. 1983. *Caligo memnon* (Buhito pardo, caligo, cream owl butterfly). In *Costa Rican natural history,* edited by D. H. Janzen. Chicago: Univ. of Chicago Press.

————. 1987. *The butterflies of Costa Rica and their natural history.* Princeton, NJ: Princeton Univ. Press.

Dietz, R. S., and J. C. Holden. 1972. The breakup of Pangaea. In *Continents adrift,* edited by J. T. Wilson. Readings from *Sci. Amer.* San Francisco: W. H. Freeman.

Dobzhansky, T. 1950. Evolution in the tropics. *Amer. Sci.* 38: 209–21.

Dressler, R. L. 1968. Pollination by euglossine bees. *Evolution* 22: 202–10.

————. 1981. *The orchids—Natural history and classification.* Cambridge, MA: Harvard Univ. Press.

Ehrlich, P. R., and P. H. Raven. 1964. Butterflies and plants: A study in coevolution. *Evolution* 18: 586–608.

————. 1967. Butterflies and plants. *Sci. Amer.* 216(6), 104–13.

Emlen, S. J., and L. W. Oring. 1977. Ecology, sexual selection, and the evolution of mating systems. *Science* 197: 215–23.

Emmons, L. H. 1984. Geographic variation in densities and diversities of non-flying mammals in Amazonia. *Biotropica* 16: 210–22.

Estrada, A., R. Coates-Estrada, and C. Vazquez-Yanes. 1984. Observations of fruiting and dispersers of *Cecropia obtusifolia* at Los Tuxtlas, Mexico. *Biotropica* 16: 315–18.

Ewel, J. 1980. Tropical succession: Manifold routes to maturity. In *Tropical succession,* supplement to *Biotropica* 12: 2–7.

————. 1983. Succession. In *Tropical rain forest ecosystems: structure and function,* edited by F. B. Golley. Amsterdam: Elsevier Scientific.

Ewel, J., C. Berish, B. Brown, N. Price, and J. Raich. 1981. Slash and burn impacts on a Costa Rican wet forest site. *Ecology* 62: 816–29.

Ewel, J., S. Gliessman, M. Amador, F. Benedict, C. Berish, R. Bermudez, B. Brown, A. Martinez, R. Miranda, and N. Price. 1982. Leaf area, light transmission, roots and leaf damage in nine tropical plant communities. *Agro-Ecosystems* 7: 305–26.

Faaborg, J. 1982. Avian population fluctuations during drought conditions in Puerto Rico. *Wilson Bull.* 94: 20–30.

Fearnside, P. M. 1986. *Human carrying capacity of the Brazilian rainforest*. New York: Columbia Univ. Press.

Fedducia, A. 1973. *Evolutionary trends in the Neotropical ovenbirds and woodhewers*. Washington, DC: American Ornithologists' Union.

Feinsinger, P. 1978. Ecological interactions between plants and hummingbirds in a successional tropical community. *Ecol. Mono.* 48: 269–87.

———. 1983. Variable nectar secretion in a *Heliconia* species pollinated by hermit hummingbirds. *Biotropica* 15: 48–52.

Feinsinger, P., and R. K. Colwell. 1978. Community organization among Neotropical nectar-feeding birds. *Amer. Zool.* 18: 779–95.

Fenton, M. B. 1983. *Just bats*. Toronto: Univ. of Toronto Press.

Fittkau, E. J., and H. Klinge. 1973. On biomass and trophic structure of the central Amazonian rain forest ecosystem. *Biotropica* 5: 1–14.

Fitzpatrick, J. W. 1980a. Foraging behavior of Neotropical tyrant flycatchers. *Condor* 82: 43–57.

———. 1980b. Wintering of North American tyrant flycatchers in the Neotropics. In *Migrant birds in the Neotropics: Ecology, behavior, distribution, and conservation*, edited by A. Keast and E. S. Morton. Washington, DC: Smithsonian Institution Press.

———. 1985. Foraging behavior and adaptive radiation in the Tyrannidae. In *Neotropical ornithology*, edited by P. A. Buckley, M. S. Foster, E. S. Morton, R. S. Ridgely, and F. G. Buckley. Washington, DC: American Ornithologists' Union.

Flannery, K. V., ed. 1982. *Maya subsistence*. New York: Academic Press.

Fleming, T. H. 1983. *Piper* (Candela, candelillos, piper). In *Costa Rican natural history*, edited by D. H. Janzen. Chicago: Univ. of Chicago Press.

———. 1985a. Coexistence of five sypatric *Piper* (Piperaceae) species in a tropical dry forest. *Ecology* 66: 688–700.

———. 1985b. A day in the life of a *Piper*-eating bat. *Nat. Hist.* 94(6): 52–59.

Forman, R.T.T. 1975. Canopy lichens with blue-green algae: A nitrogen source in a Colombian rain forest. *Ecology* 56: 1176–84.

Forshaw, J. M. 1973. *Parrots of the world*. Melbourne, Australia: Lansdowne Press.

Foster, M. S. 1977. Odd couples in manakins: A study of social organization and cooperative breeding in *Chiroxiphia linearis*. *Amer. Nat.* 111: 845–53.

Foster, R. B. 1982a. The seasonal rhythm of fruitfall on Barro Colorado Island. In *The ecology of a tropical forest*, edited by E. G.

Leigh, Jr., A. S. Rand, and D. M. Windsor. Washington, DC: Smithsonian Institution Press.

————. 1982b. Famine on Barro Colorado Island. In *The ecology of a tropical forest*, edited by E. G. Leigh, Jr., A. S. Rand, and D. M. Windsor. Washington, DC: Smithsonian Institution Press.

Furst, P. T., and M. D. Coe 1977. Ritual enemas. *Nat. Hist.* 86(3): 88–91.

Futuyma, D. J. 1983. Evolutionary interaction among herbivorous insects and plants. In *Coevolution*, edited by D. J. Futuyma and M. Slatkin. Sunderland, MA: Sinauer.

Garwood, N. C. 1982. Seasonal rhythm of seed germination in a semi-deciduous tropical forest. In *The ecology of a tropical forest*, edited by E. G. Leigh, Jr., A. S. Rand, and D. M. Windsor. Washington, DC: Smithsonian Institution Press.

Gentry, A. H. 1988. Tree species of upper Amazon forests. *Proc. Nat. Acad. Sci. U.S.A.* 85: 156–61.

Gentry, A. H., and J. L. Lopez-Parodi. 1980. Deforestation and increased flooding of the upper Amazon. *Science* 210: 1354–56.

Ghiselin, M. T. 1985. A movable feaster. *Nat. Hist.* 94(9): 54–61.

Gilbert, L. E. 1971. Butterfly-plant coevolution: Has *Passiflora adenopoda* won the selectional race with Heliconiine butterflies? *Science* 172: 585–86.

————. 1975. Ecological consequences of a coevolved mutualism between butterflies and plants. In *Coevolution of animals and plants*, edited by L. E. Gilbert and P. H. Raven. Austin, TX: Univ. of Texas Press.

————. The coevolution of a butterfly and a vine. *Sci. Amer.* 247(2): 110–21.

————. 1983. Coevolution and mimicry. In *Coevolution*, edited by D. J. Futuyma and M. Slatkin. Sunderland, MA: Sinauer.

Gladfelter, W. B., J. C. Ogden, and E. H. Gladfelter. 1980. Similarity and diversity among coral reef fish communities: Virgin Islands vs. Marshall Islands. *Ecology* 61: 1156–68.

Glander, K. E. 1977. Poison in a monkey's Garden of Eden. *Nat. Hist.* 86(3): 35–41.

Goldizen, A. W. 1988. Tamarin and marmoset mating systems: Unusual flexibility. *Trends in Ecol. Evol.* 3(2): 36–40.

Golley, F. B. 1983. Nutrient cycling and nutrient conservation. In *Tropical rain forest ecosystems: Structure and function*, edited by F. B. Golley. Amsterdam: Elsevier Scientific.

Golley, F. B., H. T. Odum, and R. F. Wilson. 1962. The structure and metabolism of a Puerto Rican red mangrove forest in May. *Ecology* 43: 9–19.

Golley, F. B., J. T. McGinnis, R. G. Clements, G. I. Child, and H. J. Duever. 1969. The structure of tropical forests in Panama and Colombia. *BioScience* 19: 693–96.

————. 1975. *Mineral cycling in a tropical moist forest ecosystem*. Athens, GA: Univ. of Georgia Press.

Gómez-Pompa, A., C. Vazquez-Yanes, and S. Guevara. 1972. The tropical rain forest: A nonrenewable resource. *Science* 177: 762–65.

Gómez-Pompa, A., H. L. Morales, E. J. Avilla, and J. J. Avilla. 1982. Experiences in traditional hyraulic agriculture. In *Maya subsistence*, edited by K. V. Flannery. New York: Academic Press.

Gotwald, W. H., Jr. 1982. Army ants. In *Social insects, vol. 4*, edited by H. R. Hermann. New York: Academic Press.

Graham, N. E., and W. B. White. 1988. The El Niño cycle: A natural oscillator of the Pacific Ocean-atmosphere system. *Science* 240: 1293–1301.

Graves, G. R., J. P. O'Neill, and T. A. Parker III. 1983. *Grallaricula ochraceifrons*, a new species of antpitta from northern Peru. *Wilson Bull.* 95: 1–6.

Greene, H. W., and R. W. McDiarmid. 1981. Coral snake mimicry: Does it occur? *Science* 213: 1207–11.

Grubb, P. J. 1977. Control of forest growth and distribution on wet tropical mountains. *Ann. Rev. Ecol. Syst.* 8: 83–107.

Haffer, J. 1969. Speciation in Amazonian forest birds. *Science* 165: 131–37.

————. 1974. *Avian speciation in tropical South America*. Publication No. 14. Cambridge, MA: Nuttall Ornithology Club.

————. 1985. Avian zoogeography of the Neotropical lowlands. In *Neotropical ornithology*, edited by P. A. Buckley, M. S. Foster, E. S. Morton, R. S. Ridgely, and F. G. Buckley. Washington, DC: American Ornithologists' Union.

Haffer, J., and J. W. Fitzpatrick. 1985. Geographic variation in some Amazonian forest birds. In *Neotropical ornithology*, edited by P. A. Buckley, M. S. Foster, E. S. Morton, R. S. Ridgely, and F. G. Buckley. Washington, DC: American Ornithologists' Union.

Hamilton, W. D. 1979. Wingless and fighting males in fig wasps and other insects. In *Reproduction, competition and selection in insects*, edited by M. S. Blum and N. A. Blum. New York: Academic Press.

Hammond, N. 1982. *Ancient Maya civilization*. New Brunswick, NJ: Rutgers Univ. Press.

Hansen, M. 1983a. Chocolate (Cacao). In *Costa Rican natural history*, edited by D. H. Janzen. Chicago: Univ. of Chicago Press.

————. 1983b. Yuca (Yuca, cassava). In *Costa Rican natural history*, edited by D. H. Janzen. Chicago: Univ. of Chicago Press.

Harborne, J. B. 1982. *Introduction to ecological biochemistry*. New York: Academic Press.

Hart, R. D. 1980. A natural ecosystem analog approach to the design of a successional crop system for tropical forest environments. In *Tropical succession*, supplement to *Biotropica* 12: 73–82.

Hartshorn, G. S. 1980. Neotropical forest dynamics. In *Tropical succession*, supplement to *Biotropica* 12: 23–30.

————. 1983a. Plants. In *Costa Rican natural history*, edited by D. H. Janzen. Chicago: Univ. of Chicago Press.

————. 1983b. *Manilkara zapota* (Nispero, chicle tree). In *Costa Rican natural history*, edited by D. H. Janzen. Chicago: Univ. of Chicago Press.

Heisler, I. L. 1983. *Nyssodesmus python* (Milpes, large forest-floor millipede). In *Costa Rican natural history*, edited by D. H. Janzen. Chicago: Univ. of Chicago Press.

Heithaus, E. R., P. A. Opler, and H. B. Baker. 1974. Bat activity and pollination of *Bauhinia pauletia*: Plant-pollinator coevolution. *Ecology* 55: 412–19.

Heithaus, E. R., T. H. Fleming, and P. A. Opler. 1975. Foraging patterns and resource utiliization in seven species of bats in a seasonal tropical forest. *Ecology* 56: 841–54.

Heyer, W. R., and L. R. Maxon. 1982. Distributions, relationships, and zoogeography of lowland frogs—The *Leptodactylus* complex in South America, with special reference to Amazonia. In *Biological diversification in the tropics*, edited by G. T. Prance. New York: Columbia Univ. Press.

Hilty, S. L., and W. L. Brown. 1986. *A guide to the birds of Colombia*. Princeton, NJ: Princeton Univ. Press.

Holdridge, L. R. 1967. *Life zone ecology*. San José, Costa Rica: Tropical Science Center.

Holthuijzen, A.M.A. and J.H.A. Boerboom. 1982. The *Cecropia* seedbank in the Surinam lowland rain forest. *Biotropica* 14: 62–67.

Horn, H. S. 1971. *The adaptive geometry of trees*. Princeton, NJ: Princeton Univ. Press.

Horn, J. M., K. C. Spencer, and J. T. Smiley. 1984. The chemistry of extrafloral nectar of *Passiflora* and related species. *Bull. Ecol. Soc. Amer.* 65: 265.

Howard, R., and A. Moore. 1980. *A complete checklist of the birds of the world*. Oxford: Oxford Univ. Press.

Howden, H. F. 1983. *Megasoma elephas* (Cornizuelo, rhinoceros beetle). In *Costa Rican natural history*, edited by D. H. Janzen. Chicago: Univ. of Chicago Press.

Howe, H. F. 1977. Bird activity and seed dispersal of a tropical wet forest tree. *Ecology* 58: 539–50.

————. 1982. Fruit production and animal activity in two tropical trees. In *The ecology of a tropical forest*, edited by E. G. Leigh, Jr., A. S. Rand, and D. M. Windsor. Washington, DC: Smithsonian Institution Press

Howe, H. F., and G. F. Estabrook. 1977. On intraspecific competition for avian dispersers in tropical trees. *Amer. Nat.* 111: 817–32.

Howell, D. J. 1976. Plant-loving bats, bat-loving plants. *Nat. Hist.* 85(2): 52–59.

Howell, T. R. 1971. An ecological study of the birds of the lowland pine savanna and adjacent rain forest in northeastern Nicaragua. *Living Bird* 10: 185–242.

Hrdy, S. B. 1981. *The woman that never evolved*. Cambridge, MA: Harvard Univ. Press.

Hubbell, S. P. 1979. Tree dispersion, abundance, and diversity in a tropical dry forest. *Science* 203: 1299–1309.

———. 1980. Seed predation and the coexistence of tree species in tropical forests. *Oikos* 35: 214–29.

Hubbell, S. P., D. F. Wiemer, and A. Adejare. 1983. An antifungal terpenoid defends a neotropical tree (*Hymenaea*) against attack by fungus-growing ants (*Atta*). *Oecologia* 60: 321–27.

Hubbell, S. P., J. J. Howard, and D. F. Wiemer. 1984. Chemical leaf repellency to an attine ant: Seasonal distribution among potential host plant species. *Ecology* 65: 1067–76.

Hubbell, S. P., and R. B. Foster. 1986a. Canopy gaps and the dynamics of a Neotropical forest. In *Plant ecology*, edited by M. J. Crawley. Oxford, England: Blackwell Scientific.

———. 1986b. Commonness and rarity in a Neotropical forest: Implications for tropical tree conservation. In *Conservation biology: The science of scarcity and diversity*, edited by M. E. Soule. Sunderland, MA: Sinauer Associates.

———. 1986c. Biology, chance, and history and the structure of tropical rain forest tree communities. In *Community ecology*, edited by J. Diamond and T. J. Case. New York: Harper & Row.

Huber, O. 1982. Significance of savanna vegetation in the Amazon territory of Venezuela. In *Biological diversification in the tropics*, edited by G. T. Prance. New York: Columbia Univ. Press.

———. 1987. Neotropical savannas: Their flora and vegetation. *Trends in Ecol. Evol.* 2: 67–71.

Idyll, C. P. 1973. The anchovy crisis. *Sci. Amer.* 228(6): 22–29.

Isler, M. L., and P. R. Isler. 1987. *The tanagers: Natural history, distribution, and identification*. Washington, DC: Smithsonian Institution Press.

Janzen, D. H. 1966. Coevolution of mutualism between ants and acacias in Central America. *Evolution* 20: 249–75.

———. 1967. Synchronization of sexual reproduction of trees within the dry season in Central America. *Evolution* 21: 620–37.

———. 1969a. Allelopathy by myrmecophytes: The ant *Azteca* as an allelopathic agent of *Cecropia*. *Ecology* 50: 147–53.

———. 1969b. Seed-eaters versus seed size, number, toxicity, and dispersal. *Evolution* 23: 1–27.

———. 1970. Herbivores and the number of tree species in a tropical forest. *Amer. Nat.* 104: 501–28.

———. 1971a. Seed predation by animals. *Ann. Rev. Ecol. Syst.* 2: 465–92.

————. 1971b. Euglossine bees as long-distance pollinators of tropical plants. *Science* 171: 203–6.

————. 1972. The uncertain future of the tropics. *Nat. Hist.* 81: 80–89.

————. 1973. Tropical agroecosystems. *Science* 182: 1212–20.

————. 1974. Tropical blackwater rivers, animals, and mast fruiting in the Dipterocarpaceae. *Biotropica* 6: 69–103.

————. 1975. *Ecology of plants in the tropics.* London: Edward Arnold.

————. 1976. Why are there so many species of insects? *Proc. XV Int. Cong. Ent.*: 84–94.

————. 1979. How to be a fig. *Ann. Rev. Ecol. Syst.* 10: 13–52.

————. 1980a. Two potential coral snake mimics in a tropical deciduous forest. *Biotropica* 12: 77–78.

————. 1980b. When is it coevolution? *Evolution* 34: 611–12.

————. 1980c. Specificity of seed-attacking beetles in a Costa Rican deciduous forest. *Journal of Ecology* 68: 929–52.

Janzen, D. H., ed. 1983a. *Costa Rican natural history.* Chicago: University of Chicago Press.

Janzen, D. H. 1983b. *Mimosa pigra* (Zarza, dormilona). In *Costa Rican natural history,* edited by D. H. Janzen. Chicago: Univ. of Chicago Press..

————. 1983c. *Brotogeris jugularis* (Perico, orange-chinned parakeet). In *Costa Rican natural history,* edited by D. H. Janzen. Chicago: Univ. of Chicago Press.

————. 1983d. *Tapirus bairdii* (Danto, danta, Baird's tapir). In *Costa Rican natural history,* edited by D. H. Janzen. Chicago: Univ. of Chicago Press.

————. 1988. Tropical ecological and biocultural restoration. *Science* 239: 243–44.

Janzen, D. H., and C. R. Carroll. 1983. *Paraponera clavata* (Bala, giant tropical ant). In *Costa Rican natural history,* edited by D. H. Janzen. Chicago: Univ. of Chicago Press.

Janzen, D. H., and C. G. Hogue. 1983. *Fulgora latenaria* (Machaca, peanuthead bug, lantern fly). In *Costa Rican natural history,* edited by D. H. Janzen. Chicago: Univ. of Chicago Press.

Janzen, D. H., and D. E. Wilson. 1983. Mammals. In *Costa Rican natural history,* edited by D. H. Janzen. Chicago: Univ. of Chicago Press.

Jordan, C. F. 1982. Amazon rain forests. *Amer. Sci.* 70: 394–401.

Jordan, C. F., F. Golley, J. D. Hall, and J. Hall. 1979. Nutrient scavenging of rainfall by the canopy of an Amazonian rain forest. *Biotropica* 12: 61–66.

Jordan, C. F., and R. Herrera. 1981. Tropical rain forests: Are nutrients really critical? *Amer. Nat.* 117: 167–80.

Jordan, C. F., and J. R. Kline. 1972. Mineral cycling: Some basic concepts and their application in a tropical rain forest. *Ann. Rev. Ecol. Syst.* 3: 33–50.

Jordano, P. 1983. Fig-seed predation and dispersal by birds. *Biotropica* 15: 38–41.

Kaplan, E. H. 1982. *A field guide to coral reefs*. Boston: Houghton Mifflin.

Karr, J. R. 1975. Production, energy pathways and community diversity in forest birds. In *Tropical ecological systems: Trends in terrestrial and aquatic research*, edited by F. B. Golley and E. Medina. New York: Springer-Verlag.

———. 1976. Seasonality, resource availability, and community diversity in tropical bird communities. *Amer. Nat.* 110: 973–94.

Karr, J. R., D. W. Schemske, and N.V.L. Brokaw. 1982. Temporal variation in the understory bird community of a tropical forest. In *The ecology of a tropical forest*, edited by E. G. Leigh, Jr., A. S. Rand, and D. M. Windsor. Washington, DC: Smithsonian Institution Press.

Karr, J. R., and K. E. Freemark. 1983. Habitat selection and environmental gradients: Dynamics in the "stable" tropics. *Ecology* 64: 1481–94.

Keast, A., and E. S. Morton, eds. 1980. *Migrant birds in the Neotropics: Ecology, behavior, distribution, and conservation*. Washington, DC: Smithsonian Institution Press.

Keeler, K. H. 1980. Distribution of plants with extrafloral nectaries in temperate communities. *Amer. Midl. Nat.* 104: 274–80.

Kettlewell, B. 1973. *The evolution of melanisms: The study of a recurring necessity with special industrial melanism in the Lepidoptera*. Oxford, England: Oxford Univ. Press.

Kiltie, R. A. 1982. Bite force as a basis for niche differentiation between rain forest peccaries (*Tayassu tajacu* and *T. pecari*). *Biotropica* 14: 188–95.

Kinzey, W. G. 1982. Distribution of primates and forest refuges. In *Biological diversification in the tropics*, edited by G. T. Prance. New York: Columbia Univ. Press.

Klinge, H., W. A. Rodrigues, E. Brunig, and E. J. Fittkau. 1975. Biomass and structure in a central Amazonian rain forest. In *Tropical ecological systems: Trends in terrestrial and aquatic research*, edited by F. B. Golley and E. Medina. New York: Springer-Verlag.

Knight, D. H. 1975. A phytosociological analysis of species rich tropical forest on Barro Colorado Island, Panama. *Ecol. Mono.* 45: 259–84.

Koford, C. B. 1957. The vicuna and the pina. *Ecol. Mono.* 27: 153–219.

Krantz, G. S. 1970. Human activities and megafaunal extinctions. *Amer. Sci.* 58: 164–70.

Kricher, J. C., and W. E. Davis, Jr. 1986. Returns and winter site fidelity of North American migrants banded in Belize, Central America. *J. Field Ornithol.* 57: 48–52.

———. 1987. No place like home. *Living Bird Quarterly* 6(3): 24–27.

Krieger, R. I., L. P. Feeny, and C. F. Wilkinson. 1971. Detoxication enzymes in the guts of caterpillars: An evolutionary answer to plant defenses? *Science* 172: 579–81.

Kunz, T. H. 1982. Roosting ecology of bats. In *Ecology of bats*, edited by T. H. Kunz. New York: Plenum Publishing Corp.

Kunz, T. H., P. V. August, and C. D. Burnett. 1983. Harem social organization in cave roosting *Artibeus jamaicensis* (Chiroptera: Phyllostomidae). *Biotropica* 15: 133–38.

Kushlan, J. A., G. Morales, and P. C. Frohring. 1985. Foraging niche relations of wading birds in tropical wet savannas. In *Neotropical ornithology*, edited by P. A. Buckley, M. S. Foster, E. S. Morton, R. S. Ridgely, and F. G. Buckley. Washington, DC: American Ornithologists' Union.

Lack, D. 1947. *Darwin's finches*. Cambridge, MA: Cambridge Univ. Press.

———. 1966. *Population studies of birds*. Oxford, England: Clarendon Press.

LaFay, H. 1975. The Maya: Children of time. *Nat. Geog.* 148: 728–67.

Land, H. 1970. *Birds of Guatemala*. Wynnewood, PA: Livingston.

Leck, C. F. 1969. Observations of birds exploiting a Central American fruit tree. *Wilson Bull.* 81: 264–69.

———. 1972. Seasonal changes in feeding pressures of fruit and nectar-eating birds in Panama. *Condor* 74: 54–60.

———. 1987. Habitat selection in migrant birds: Seductive fruits. *Trends in Ecol. Evol.* 2(2): 33.

Leigh, E. G., Jr. 1975. Structure and climate in tropical rain forest. *Ann. Rev. Ecol. Syst.* 6: 67–86.

Leigh, E. G. Jr., A. S. Rand, and D. M. Windsor, eds. 1982. *The ecology of a tropical forest*. Washington, DC: Smithsonian Institution Press.

Leigh, E. G., Jr., and D. M. Windsor. 1982. Forest production and regulation of primary consumers on Barro Colorado Island. In *The ecology of a tropical forest*, edited by E. G. Leigh, Jr., A. S. Rand, and D. M. Windsor. Washington, DC: Smithsonian Institution Press.

Levey, D. J. 1985. Two ways to be a fruit-eating bird: Mashers versus gulpers. Abstracts of the 103d American Ornithologists' Union Meeting.

Levey, D. J., T. C. Moermond, and J. S. Denslow. 1984. Fruit choice in Neotropical birds: The effect of distance between fruits on preference patterns. *Ecology* 65: 844–50.

Levin, D. A. 1971. Plant phenolics: An ecological perspective. *Amer. Nat.* 105: 157–81.

———. 1976. Alkaloid-bearing plants: An ecogeographic perspective. *Amer. Nat.* 110: 261–84.

Levings, S. C., and D. M. Windsor. 1982. Seasonal and annual variation in litter arthropod populations. In *The ecology of a tropical forest,* edited by E. G. Leigh, Jr., A. S. Rand, and D. M. Windsor. Washington, DC: Smithsonian Institution Press.

Lill, A. 1974. The evolution of clutch size and male "chauvinism" in the white-bearded manakin. *Living Bird* 13: 211–31.

Limbaugh, C. 1961. Cleaning symbiosis. *Sci. Amer.* 205(2): 42–49.

Long, A., and P. S. Martin. 1974. Death of American ground sloths. *Science* 186: 638–40.

Longman, K. A., and J. Jenik. 1974. *Tropical forest and its environment.* London: Longman.

Lopez-Parodi, J. 1980. Deforestation and increased flooding of the upper Amazon. *Science* 210: 1354–55.

Lovejoy, T. E. 1974. Bird diversity and abundance in Amazon forest communities. *Living Bird* 13: 127–92.

Lowe-McConnell, R. H. 1987. *Ecological studies in tropical fish communities.* Cambridge: Cambridge Univ. Press.

Lubin, Y. D. 1983. *Nasutitermes* (Comejan, hormiga blanca, nausute termite, arboreal termite). In *Costa Rican natural history,* edited by D. H. Janzen. Chicago: Univ. of Chicago Press.

Lubin, Y. D., and G. G. Montgomery. 1981. Defenses of *Nasutitermes* termites (Isoptera, Termitidae) against *Tamandua* anteaters (Edentata, Myrmecophagidae). *Biotropica* 13: 66–76.

Lugo, A. E., and S. C. Snedaker. 1974. The ecology of mangroves. *Ann. Rev. Ecol. Syst.* 5: 39–64.

MacArthur, R. H. 1965. Patterns of species diversity. *Biological Review* 40: 510–33.

———. 1972. *Geographical ecology: Patterns in the distribution of species.* New York: Harper & Row.

MacArthur, R. H., and J. MacArthur. 1961. On bird species diversity. *Ecology* 42: 594–98.

Mares, M. A. 1986. Conservation in South America: Problems, consequences, and solutions. *Science* 233: 734–39.

Markell, E. K., and M. Voge. 1971. *Medical parasitology.* 3d ed. Philadelphia: Saunders.

Marquis, R. J. 1984. Leaf herbivores decrease fitness of a tropical plant. *Science* 226: 537–39.

Martin, M. M. 1970. The biochemical basis of the fungus-attine ant symbiosis. *Science* 169: 16–20.

Martin, M. M., and J. S. Martin. 1984. Surfactants: Their role in preventing the precipitation of proteins by tannins in insect guts. *Oecologia* 61: 342–345.

Maxson, L. R., and W. R. Heyer. 1982. Leptodactylid frogs and the Brasilian Shield: An old and continuing adaptive relationship. *Biotropica* 14: 10–14.

Maxson, L. R., and C. W. Myers. 1985. Albumin evolution intropical poison frogs (Dendrobatidae): A preliminary report. *Biotropica* 17: 50–56.

Mayr, E. 1963. *Animal species and evolution.* Cambridge, MA: Belknap Press, Harvard.

McLarney, W. O. 1988. Guanacaste: The dawn of a park. *Nature Conservancy Magazine* 38(1): 11–15.

Meyer-Abich, Adolf. 1969. *Alexander von Humboldt.* Bonn: Inter Nationes.

Meyer de Schauensee, R. 1966. *The species of birds of South America.* Wynnewood, PA: Livingston, for the Academy of Natural Sciences, Philadelphia.

Milton, K. 1979. Factors influencing leaf choice by howler monkeys: A test of some hypotheses of food selection by generalist herbivores. *Amer. Nat.* 114: 362–78.

———. 1981. Food choice and digestive strategies of two sympatric primate species. *Amer Nat.* 117: 496–505.

———. 1982. Dietary quality and demographic regulation in a howler monkey population. In *The ecology of a tropical forest,* edited by E. G. Leigh, Jr., A. S. Rand, and D. M. Windsor. Washington, DC: Smithsonian Institution Press.

Moermond, T. C., and J. S. Denslow. 1985. Neotropical avian frugivores: Patterns of behavior, morphology, and nutrition, with consequences for fruit selection. In *Neotropical ornithology,* edited by P. A Buckley, M. S. Foster, E. S. Morton, R. S. Ridgely, and F. G. Buckley. Washington, DC: American Ornithologists' Union.

Montgomery, G. G., and M. E. Sunquist. 1975. Impact of sloths on Neotropical forest energy flow and nutrient cycling. In *Ecological studies 11, Tropical ecological systems,* edited by F. B. Golley and E. Medina. Heidelberg: Springer-Verlag.

Morrison, T. 1974. *Land above the clouds.* London: Andre Deutsch.

———. 1976. *The Andes.* Amsterdam: Time-Life International.

Morton, E. S. 1973. On the evolutionary advantages and disadvantages of fruit eating in tropical birds. *Amer. Nat.* 107: 8–22.

Moynihan, M. 1962. The organization and probable evolution of some mixed species flocks of Neotropical birds. *Smithsonian Miscellaneous Collection* 143: 1–140.

———. 1976. *The new world primates: Adaptive radiation and the evolution of social behavior, languages, and intelligence.* Princeton, NJ: Princeton Univ. Press.

Müller, F. 1879. *Ituna* and *Thyridis*: A remarkable case of mimicry in butterflies. *Proc. Ent. Soc. London 1879*: 20–29.

Munn, C. A. 1984. Birds of different feather also flock together. *Nat. Hist.* 11(11): 34–42.

———. 1985. Permanent canopy and understory flocks in Amazonia: Species composition and population density. In *Neotropical ornithology,* edited by P. A. Buckley, M. S. Foster, E. S. Morton, R. S. Ridgely, and F. G. Buckley. Washington, DC: American Ornithologists' Union.

Munn, C. A., and J. W. Terborgh. 1979. Multi-species territoriality in Neotropical foraging flocks. *Condor* 81: 338–47.

Myers, N., for the Committee on Research Priorities in Tropical Biology of the National Research Council. 1980. *Conversion of tropical moist forests.* Washington, DC: National Academy of Science.

————. 1984. *The primary source: Tropical forests and our future.* New York: W. W. Norton.

Nadkarni, N. M. 1981. Canopy roots: convergent evolution in rainforest nutrient cycles. *Science* 214: 1023–24.

Nathanson, J. A. 1984. Caffeine and related methylxanthines: Possible naturally occurring pesticides. *Science* 226: 184–86.

Norris, R. 1988. Data for diversity. *Nature Conservancy Magazine* 38(1): 4–10.

Oliveira, P. S., and H. F. Leitao-Filho. 1987. Extrafloral nectaries: Their taxonomic distribution and abundance in the woody flora of cerrado vegetation in southeast Brazil. *Biotropica* 19: 140–48.

O'Neill, J. P., and G. R. Graves. 1977. A new genus and species of owl (Aves: Strigidae) from Peru. *Auk* 94: 409–16.

Onuf, C. P., J. M. Teal, and I. Valiela. 1977. Interactions of nutrients, plant growth, and herbivory in a mangrove ecosystem. *Ecology* 58: 514–26.

Opler, P. A. 1981. Polymorphic mimicry by a neuropteran. *Biotropica* 13: 165–76.

Opler, P. A., H. G. Baker, and G. W. Frankie. 1980. Plant reproductive characteristics during secondary succession in Neotropical lowland forest ecosystems. In *Tropical succession,* supplement to *Biotropica* 12: 40–46.

Oppenheimer, J. R. 1982. *Cebus capucinus*: home range, population dynamics, and interspecific relationships. In *The ecology of a tropical forest,* edited by E. G. Leigh, Jr., A. S. Rand, and D. M. Windsor. Washington, DC: Smithsonian Institution Press.

Oster, G., and S. Oster. 1985. The great breadfruit scheme. *Nat. Hist.* 94: 35–41.

Otte, D. 1980. Beetles adorned with horns. *Nat. Hist.* 89: 34–41.

Papageorgis, C. 1975. Mimicry in Neotropical butterflies. *Amer. Sci.* 63: 522–32.

Parker, T. A., III, and J. P. O'Neill. 1985. A new species and new subspecies of *Thryothorus* wren from Peru. In *Neotropical Ornithology,* edited by P. A. Buckley, M. S. Foster, E. S. Morton, R. S. Ridgely, and F. G. Buckley. Washington, DC: American Ornithologists' Union.

Pearson, D. L. 1977. A pantropical comparison of bird community structure on six lowland forest sites. *Condor* 79: 232–44.

————. 1982. Historical factors and bird species richness. In *Biological diversification in the tropics,* edited by G. T. Prance. New York: Columbia Univ. Press.

Pearson, D. L., and J. A. Derr. 1986. Seasonal patterns in lowland forest floor arthropod abundance in southeastern Peru. *Biotropica* 18: 244–56.

Perrins, C. M., and A. L. A. Middleton, eds. 1985. *The encyclopedia of birds*. New York: Facts on File.

Perry, D. R. 1978. Factors influencing arboreal epiphytic phytosociology in Central America. *Biotropica* 10: 235–37.

———. 1984. The canopy of the tropical rain forest. *Sci. Amer.* 251(5): 138–47.

Peterson, R. T., and E. L. Chaliff. 1973. *A field guide to Mexican birds*. Boston: Houghton Mifflin.

Pianka, E. R. 1966. Latitudinal gradients in species diversity: A review of concepts. *Amer. Nat.* 100: 33–45.

Picado, C. 1913. Les bromeliacees epiphytes considerees comme milieu biologique. *Bull. Sci. France Belgique ser.* 7 47: 216–360.

Porter, J. W. 1972. Patterns of species diversity in Caribbean reef corals. *Ecology* 53: 745–48.

———. 1976. Autotrophy, heterotrophy, and resource partitioning in Caribbean reef-building corals. *Amer. Nat.* 110: 731–42.

Pough, F. H. 1973. Lizard energetics and diet. *Ecology* 54. 837 44

Prance, G. T., ed. 1982a. *Biological diversification in the tropics*. New York: Columbia Univ. Press.

Prance, G. T. 1982b. Forest refuges: Evidence from woody angiosperms. In *Biological diversification in the tropics*, edited by G. T. Prance. New York: Columbia Univ. Press.

Putz, F. E. 1984. The natural history of lianas on Barro Colorado Island, Panama. *Ecology* 65: 1713–24.

Rappole, J. H., Morton, F. S., Lovejoy, T. E., III, and J. L. Ruos. 1983. *Nearctic avian migrants in the Neotropics*. Washington, DC: U. S. Department of the Interior, Fish and Wildlife Service.

Rathcke, B. J., and R. W. Poole. 1975. Coevolutionary race continues: Butterfly larval adaptation to plant trichomes. *Science* 187: 175–76.

Raven, P. H., and D. Axelrod. 1975. History of the flora and fauna of Latin America. *Amer. Sci.* 63: 420–29.

Ray, T. S., and C. C. Andrews. 1980. Antbutterflies: Butterflies that follow army ants to feed on antbird droppings. *Science* 210: 1147–48.

Remsen, J. V., Jr. 1984. High incidence of "leapfrog" pattern of geographic variation in Andean birds: Implications for the speciation process. *Science* 224: 171–72.

Remsen, J. V., Jr. and T. A. Parker III. 1983. Contribution of river-created habitats to bird species richness in Amazonia. *Biotropica* 15: 223–31.

Remsen, J. V., Jr. and T. A. Parker III. 1985. Bamboo specialists among Neotropical birds. Abstracts of the 103d American Ornithologists' Union Meeting.

Rettenmeyer, C. W. 1983. *Eciton burchilli* and other army ants (Hormiga arriera, army ants). In *Costa Rican natural history*, edited by D. H. Janzen. Chicago: Univ. of Chicago Press.

Richards, P. W. 1952. *The tropical rain forest*. Cambridge: Cambridge Univ. Press.

Richards, P. W. 1973. The tropical rain forest. *Sci. Amer.* 229(6): 58–67.

Ricklefs, R. E. 1969a. The nesting cycle of songbirds in tropical and temperate regions. *Living Bird* 8: 165–75.

——. 1969b. Natural selection and the development of mortality rates in young birds. *Nature* 223: 922–25.

——. 1970. Clutch size in birds: Outcome of opposing predator and prey adaptations. *Science* 168: 599–600.

——. 1977. Environmental heterogeniety and plant species diversity: A hypothesis. *Amer. Nat.* 111: 376–81.

Ridgely, R. S. 1976. *A guide to the birds of Panama.* Princeton, NJ: Princeton Univ. Press.

Robinson, S. K. 1985a. Coloniality in the yellow-rumped cacique as a defense against nest predators. *Auk* 102: 506–19.

——. 1985b. The yellow-rumped cacique and its associated nest pirates. In *Neotropical ornithology,* edited by P. A. Buckley, M. S. Foster, E. S. Morton, R. S. Ridgely, and F. G. Buckley. Washington, DC: American Ornithologists' Union.

——. 1986. Social security for birds. *Nat. Hist.* 95(3): 39–47.

Rockwood, L. L. 1976. Plant selection and foraging patterns in two species of leaf-cutting ants (*Atta*). *Ecology* 57: 48–61.

Rodriguez, G. 1987. Structure and production in Neotropical mangroves. *Trends in Ecol. Evol.*: 2(9) 264–67.

Rosenberg, G. H. 1985. Birds specialized on Amazon river islands. Abstracts of the 103d American Ornithologists' Union Meeting.

Ryan, C. A. 1979. Proteinase inhibitors. In *Herbivores: Their interaction with secondary plant metabolites,* edited by G. A. Rosenthal and D. H. Janzen. New York: Academic Press.

Salati, E., and P. B. Vose. 1984. Amazon Basin: A system in equilibrium. *Science* 225: 129–38.

Salick, J., R. Herrera, and C. F. Jordan. 1983. Termitaria: Nutrient patchiness in nutrient-deficient rain forests. *Biotropica* 15: 1–7.

Sanford, R. L., Jr. 1987. Apogeotropic roots in an Amazon rain forest. *Science* 235: 1062–64.

Sarmiento, G. 1983. The savannas of Tropical America. In *Tropical savannas,* edited by F. Boulière. New York: Elsevier.

Sarmiento, G., and M. Monasterio. 1975. A critical consideration of the environmental conditions associated with the occurrence of savanna ecosystems in tropical America. In *Tropical ecological systems: Trends in terrestrial and aquatic research,* edited by F. B. Golley and E. Medina. New York: Springer-Verlag.

Schoener, T. W. 1971. Large-billed insectivorous birds: A precipitous diversity gradient. *Condor* 73: 154–61.

Schulenberg, T. S., and M. D. Williams. 1982. A new species of antpitta (*Grallaria*) from northern Peru. *Wilson Bull.* 94: 105–13.

Schwartz, P., and D. W. Snow. 1978. Display and related behavior of the wire-tailed manakin. *Living Bird* 17: 51–78.

Scott, N. J., and S. Limerick. 1983. Reptiles and amphibians. In *Costa Rican natural history*, edited by D. H. Janzen. Chicago: Univ. of Chicago Press.

Scott, P. E., and R. F. Martin. 1983. Reproduction of the turquoise-browed motmot at archaeological ruins in Yucatan. *Biotropica* 15: 8–14.

Seilger, D., and P. W. Price. 1976. Secondary compounds in plants: Primary functions. *Amer. Sci.* 110: 101–5.

Sherry, T. W. 1984. Comparative dietary ecology of sympatric, insectivorous neotropical flycatchers. *Ecol. Mono.* 54: 313–38.

Short, L. L. 1982. *Woodpeckers of the world*. Greenville, DE: Delaware Museum of Natural History.

Sibley, C. G., and J. E. Ahlquist. 1983. Phylogeny and classification of birds based on the data of DNA-DNA hybridization. In *Current ornithology*, edited by R. F. Johnston. New York: Plenum Press.

Sick, H. 1967. Courtship behavior in manakins (Pipridae): A review. *Living Bird* 6: 5–22.

Siemans, A. H. 1982. Prehispanic agricultural use of the wetlands of northern Belize. In *Maya subsistence*, edited by F. V. Fleming. New York: Academic Press.

Simpson, B. B., and J. Haffer. 1978. Speciation patterns in the Amazonian forest biota. *Ann. Rev. Ecol. Syst.* 9: 497–518.

Skutch, A. F. 1954. *Life histories of Central American birds*. Pacific Coast Avifauna No. 31. Berkeley, CA: Cooper Ornithological Society.

———. 1960. *Life histories of Central American birds II*. Pacific Coast Avifauna No. 34. Berkeley, CA: Cooper Ornithological Society.

———. 1967. *Life histories of Central American highland birds*. Publication No. 7. Cambridge, MA: Nuttall Ornithology Club.

———. 1969. *Life histories of Central American birds III*. Pacific Coast Avifauna No. 35. Berkeley, CA: Cooper Ornithological Society.

———. 1972. *Studies of tropical American birds*. Publication No. 10. Cambridge, MA: Nuttall Ornithology Club.

———. 1973. *The life of the hummingbird*. New York: Crown

———. 1981. *New studies of tropical American birds*. Publication No. 19. Cambridge, MA: Nuttall Ornithology Club.

———. 1983. *Birds of Tropical America*. Austin, TX: Univ. of Texas Press.

———. 1985. *The life of the woodpecker*. Santa Monica, CA: Ibis.

Smiley, J. T. 1985. *Heliconius* caterpillar mortality during establishment on plants with and without attending ants. *Ecology* 66: 845–49.

Smith, N. J. H. 1981. Colonization lessons from a tropical forest. *Science* 214: 755–60.

Smith, S. M. 1975. Innate recognition of coral snake pattern by a possible avian predator. *Science* 187: 759–60.

———. 1977. Coral snake pattern rejection and stimulus generalisation by naive great kiskadees (Aves: Tyrannidae). *Nature* 265: 535–36.

Smythe, N., W. E. Glanz, and E. G. Leigh, Jr. 1982. Population regulation in some terrestrial frugivores. In *The ecology of a tropical forest*, edited by E. G. Leigh, Jr., A. S. Rand, and D. M. Windsor. Washington, DC: Smithsonian Institution Press.

Snow, B. K., and D. W. Snow. 1979. The ochre-bellied flycatcher and the evolution of lek behavior. *Condor* 81: 286–92.

Snow, D. W. 1961. The natural history of the oilbird, *Steatornis caripensis*, in Trinidad, W. I. Part 1. General behavior and breeding habits. *Zoologica* 46: 27–48.

———. 1962a. A field study of the black-and-white manakin, *Manacus manacus*, Trinidad. *Zoologica* 47: 65–104.

———. 1962b. The natural history of the oilbird, *Steatornis caripensis*, in Trinidad, W. I. Part 2. Population, breeding ecology, food. *Zoologica* 47: 199–221.

———. 1966. A possible selective factor in the evolution of fruiting seasons in tropical forest. *Oikos* 15: 274–81.

———. 1971. Observations on the purple-throated fruit-crow in Guyana. *Living Bird* 10: 5–18.

———. 1976. *The web of adaptation*. New York: Demeter Press-Quadrangle.

———. 1982. *The cotingas: Bellbirds, umbrellabirds, and other species*. Ithaca, NY: Cornell Univ. Press.

Sowls, L. K. 1984. *The peccaries*. Tucson, AZ: Univ. of Arizona Press.

Spencer, K. C. 1984. Chemical correlates of coevolution: The *Passiflora/Heliconius* interaction. *Bull. Ecol. Soc. Amer.* 65: 231.

Stiles, E. W. 1980. Patterns of fruit presentation and seed dispersal in bird-disseminated woody plants in the eastern deciduous forest. *Amer. Nat.* 116: 670–88.

———. 1984. Fruit for all seasons. *Nat. Hist* 93(8): 42–53.

Stiles, F. G. 1975. Ecology, flowering phenology and hummingbird pollination of some Costa Rican *Heliconia* species. *Ecology* 56: 285–301.

———. 1977. Coadapted competitors: The flowering seasons of hummingbird-pollinated plants in a tropical forest. *Science* 198: 1177–78.

———. 1983. *Heliconia latispatha* (Platanillo, wild plantain). In *Costa Rican natural history*, edited by D. H. Janzen. Chicago: Univ. of Chicago Press.

Stoddard, D. R. 1969. Post-hurricane changes in the British Honduras reefs and cays. *Atoll Research Bulletin* 131: 1–25.

Storer, R. W. 1969. What is a tanager? *Living Bird* 8: 127–36.

Stradling, D. J. 1978. The influence of size on foraging in the ant *Atta cephalotes*, and the effect of some plant defense mechanisms. *Jour. Animal Ecology* 47: 173–88.

Strahl, S. D. 1985. Correlates of reproductive success in communal

Hoatzins (*Opisthocomus hoazin*). Abstracts of the 103d American Ornithologists' Union Meeting.

Terborgh, J., and J. S. Weske. 1975. The role of competition in the distribution of Andean birds. *Ecology* 56: 562–76.

Toledo, V. M. 1977. Pollination of some rainforest plants by non-hovering birds in Veracruz, Mexico. *Biotropica* 9: 262–67.

Trail, P. W. 1985a. Courtship disruption modifies mate choice in a lek-breeding bird. *Science* 227: 778–79.

———. 1985b. A lek's icon: The courtship display of a Guianan cock-of-the-rock. *Amer. Birds* 39: 235–40.

Tramer, E. J. 1974. On latitudinal gradients in avian diversity. *Condor* 76: 123–30.

Traylor, M. A., Jr., and J. W. Fitzpatrick. 1982. A survey of the tyrant flycatchers. *Living Bird* 19: 7–50.

Truell, P. 1988. What do monkeys in Bolivia have to do with the debt crisis? *Wall Street Journal*, January 20, 1988.

Turner, B. L., II, and P. D. Harrison. 1981. Prehistoric raised-field agriculture in the Maya lowlands. *Science* 213: 339–405.

Turner, D. C. 1975. *The vampire bat: A field study in behavior and ecology*. Baltimore, MD: Johns Hopkins Univ. Press.

Turner, J.R.G. 1971. Studies of Müllerian mimicry and its evolution in burnet moths and heliconid butterflies. In *Ecological genetics and evolution*, edited by R. Creed. Oxford, England: Blackwell Scientific.

———. 1975. A tale of two butterflies. *Nat. Hist.* 84(2): 29–37.

———. 1981. Adaptation and evolution in *Heliconius*: A defense of neo-Darwinism. *Ann. Rev. Ecol. Syst.* 12: 99–121.

Tuttle, M. D., and M. J. Ryan. 1981. Bat predation and the evolution of frog vocalizations in the neotropics. *Science* 214. 677–78.

Utley, J. F., and K. Burt-Utley. 1983. Bromeliads (Pina silvestre, pinuelas, chiras, wild pineapple). In *Costa Rican natural history*, edited by D. H. Janzen. Chicago: Univ. of Chicago Press.

Vandermeer, J. 1983a. African oil palm (Palma de aceite). In *Costa Rican natural history*, edited by D. H. Janzen. Chicago: Univ. of Chicago Press.

———. 1983b. Banana (Platano, banano). In *Costa Rican natural history*, edited by D. H. Janzen. Chicago: Univ. of Chicago Press.

———. 1983c. Coconut (Coco). In *Costa Rican natural history*, edited by D. H. Janzen. Chicago: Univ. of Chicago Press.

Van Devender, R. W. 1983. *Basiliscus basiliscus* (Chisbala, garrobo, basilisk, Jesus Christ lizard). In *Costa Rican natural history*, edited by D. H. Janzen. Chicago: Univ. of Chicago Press.

Vickers, W. T. 1988. Game depletion hypothesis of Amazonian adaptation: Data from a native community. *Science* 239: 1521–22.

Wahle, C. M. 1980. Detection, pursuit, and overgrowth of tropical gorgonians by milleporid hydrocoarals: Perseus and Medusa revisited. *Science* 209: 689–91.

Wallace, A. R. 1876. The geographical distribution of animals. With a study of the relations of living and extinct faunas as elucidating the past changes in the earth's surface. New York: Harper & Bros.

———. 1895. *Natural selection and tropical nature*. London: Macmillan.

Walsh, G. E. 1974. Mangroves: A review. In *Ecology of Halophytes*, edited by R. J. Reimold and W. H. Queens. New York: Academic Press.

Walter, H. 1971. *Ecology of tropical and subtropical vegetation*. New York: Van Nostrand Reinhold.

———. 1973. *Vegetation of the earth in relation to climate and the ecophysiological conditions*. London: English Universities Press.

Walterm, K. S. 1983. Orchidaceae (Orquideas, orchids). In *Costa Rican natural history*, edited by D. H. Janzen. Chicago: Univ. of Chicago Press.

Waterton, C. [1825]. 1983. *Wanderings in South America*. Reprint. London: Century Publishing.

Weber, N. A. 1972. The attines: The fungus-culturing ants. *Amer. Sci.* 60: 448–56.

Wheelwright, N. T. 1985. Fruit size, gape width, and the diets of fruit-eating birds. *Ecology* 66: 808–18.

Wheelwright, N. T., W. A. Haber, K. G. Murray, and C. Guindon. 1984. Tropical fruit-eating birds and their food plants: A survey of a Costa Rican lower Montane forest. *Biotropica* 16: 173–92.

Whittaker, R. H. 1975. *Communities and ecosystems*. 2d ed. New York: Macmillan.

Whittaker, R. H., and P. P. Feeny. 1971. Allelochemics: Chemical interactions between species. *Science* 171: 757–70.

Wiebes, J. T. 1979. Co-evolution of figs and their insect pollinators. *Ann. Rev. Ecol. Syst.* 10: 1–12.

Williams-Linera, G. 1983. Biomass and nutrient content in two successional stages of tropical wet forest in Uxpanapa, Mexico. *Biotropica* 15: 275–84.

Willis, E. O. 1966. The role of migrant birds at swarms of army ants. *Living Bird* 5: 187–232.

———. 1967. The behavior of the bicolored ant birds. *University of California Publ. Zool.* 79: 1–127.

Willis, E. O., and Y. Oniki. 1978. Birds and army ants. *Ann. Rev. Ecol. Syst.* 9: 243–63.

Wilson, E. O. 1971. *The insect societies*. Cambridge, MA: Belknap Press, Harvard.

———. 1978. *On human nature*. Cambridge, MA: Harvard Univ. Press.

———. 1987. The arboreal ant fauna of Peruvian Amazon forests: A first assessment. *Biotropica* 19: 245–51.

Wolf, L. L. 1975. "Prostitution" behavior in a tropical hummingbird. *Condor* 77: 140–44.

Woodley, J. D., E. A. Chornesky, P. A. Clifford, J.B.C. Jackson, L. S. Kaufman, N. Knowlton, J. C. Lang, M. P. Pearson, J. W. Porter, M. C. Rooney, K. W. Rylaarsdam, V. J. Tunnicliffe, C. M. Wahle, J. L. Wulff, A.S.G. Curtis, M. D. Dallmeyer, B. P. Jupp, M.A.R. Koehl, J. Neigel, and E. M. Sides. 1981. Hurricane Allen's impact on Jamaican coral reefs. *Science* 214: 749–61.

Woodwell, G. H. 1978. The carbon dioxide question. *Sci. Amer.* 238(1): 34–43.

Worthington, A. 1982. Population sizes and breeding rhythms of two species of manakins in relation to food supply. 1982. In *The ecology of a tropical forest*, edited by E. G. Leigh, Jr., A. S. Rand, and D. M. Windsor. Washington, DC: Smithsonian Institution Press.

Wunderle, J. M., Jr. 1982. The timing of the breeding season in the Bananaquit (Coereba flaveola) on the island of Grenada, W.I. *Biotropica* 14: 124–31.

Young, A. M. 1971. Wing coloration and reflectance in *Morpho* butterflies as related to reproductive behavior and escape from avian predators. *Oecologia* 7: 209–22.

Zimmerman, P. R., J. P. Greenberg, S. O. Wandiga, and P. J. Crutzen. 1982. Termites: A potentially large source of atmospheric methane, carbon dioxide, and molecular hydrogen. *Science* 218: 563–65.

Zucker, W. V. 1983. Tannins. Does structure determine function? An ecological perspective. *Amer. Nat.* 121: 355–65.

FIELD GUIDES TO NEOTROPICAL BIRDS

Alden, P. 1969. *Finding the birds in western Mexico.* Tucson, AZ: Univ. of Arizona Press.

Davis, L. I. 1972. *A field guide to the birds of Mexico and Central America.* Austin, TX: Univ. of Texas.

de Schauensee, R. M., and W. H. Phelps, Jr. 1978. *A guide to the birds of Venezuela.* Princeton, NJ: Princeton Univ. Press.

Dunning, J. S. 1982. *South American land birds: A photographic aid to identification.* Newtown Square, PA: Harrowood Books.

Edwards, E. P. 1972. *A field guide to the birds of Mexico.* Sweet Briar, VA: E. P. Edwards.

ffrench, R. 1976. *A guide to the birds of Trinidad and Tobago.* Valley Forge, PA: Harrowood.

Frisch, J. D. 1981. *Aves Brasileiras.* San Paulo: Dalgas-Ecoltec Ecologia Tecnica e Comercio Ltda.

Hilty, S. L., and W. L. Brown. 1986. *A guide to the birds of Colombia.* Princeton, NJ: Princeton Univ. Press.

Koepcke, M. 1964. *The birds of the department of Lima, Peru.* Wynnewood, PA: Livingston.

Land, H. C. 1970. *Birds of Guatemala.* Wynnewood, PA: Livingston.

Peterson, R. T. and E. L. Chaliff. 1973. *A field guide to Mexican birds.* Boston: Houghton Mifflin.

Raffaele, H. A. 1983. *A guide to the birds of Puerto Rico and the Virgin Islands.* San Juan: Fondo Educativo Interamericano.

Ridgely, R. S. 1976. *A guide to the birds of Panama.* Princeton, NJ: Princeton Univ. Press.

———. 1989. *A guide to the birds of Panama.* 2d ed. Princeton, NJ: Princeton Univ. Press. In press.

Smithe, F. D. 1966. *The birds of Tikal.* New York: Natural History Press.

GENERAL BOOK-LENGTH REFERENCES

The following books were useful to me in preparing this volume. I recommend them strongly to the reader who wishes to delve further into the literature about tropical biology. Some are written for the general reader, some are technical, and some are classic accounts by the early explorer-naturalists. I note these distinctions in my comments about each book.

Bates, H. W. 1892. *The naturalist on the river amazons.* London: John Murray. Classic account of Amazonian natural history.

Bates, M. 1964. *The land and wildlife of South America.* New York: Time Inc. Popular account of the American tropics, with excellent illustrations.

Beebe, W. 1918. *Jungle peace.* London: Witherby.

Beebe, W. 1921. *Edge of the jungle.* New York: Henry Holt. Both this and the previous volume contain short, delightful essays on tropical ecology. Classics.

Belt, T. [1874] 1985. *The naturalist in Nicaragua.* Chicago: Univ. of Chicago Press. One of the best of the classic exploratory accounts.

Buckley, P. A., M. S. Foster, E. S. Morton, R. S. Ridgely, and F. G. Buckley. 1985. *Neotropical ornithology.* Washington, DC: American Ornithologists' Union. This large volume contains sixty three technical papers about tropical bird ecology.

Caufield, C. 1984. *In the rainforest.* Chicago: Univ. of Chicago Press. Discusses both tropical ecology and conservation issues for the general reader.

Chapman, F. M. 1938. *Life in an air castle.* New York: Appleton-Century. A classic account, highly readable with much information.

Committee on Research Priorities in Tropical Biology of the National Research Council. 1980. *Research priorities in tropical biology.* Washington, DC: National Academy of Science. Outlines basic questions about tropical ecosystems as yet not fully answered. Oriented toward concerns of conservationists.

Darwin, C. R. [1906] 1959. *The voyage of the Beagle.* Reprint. London: J. M. Dent and Sons. One of the best accounts of travel throughout South America.

Eisenberg, J. F., ed. 1979. *Vertebrate ecology in the northern Neotropics.* Washington, DC: Smithsonian Institution Press. Deals mostly with mammals, little on birds or other vertebrates. Good, but technical.

Emmel, T. C. 1976. *Population biology.* New York: Harper & Row. Contains a well-written chapter on tropical ecology, especially seasonality. Very oriented to adaptations of butterflies.

Ewel, J., ed. 1980. *Tropical succession*, supplement to *Biotropica* 12. A collection of thirteen technical papers on the topic of succession. Several were discussed in this book.

Forsyth, A., and K. Miyata. 1984. *Tropical nature*. New York: Charles Scribner's Sons. A very readable series of essays on Neotropical ecology for the general reader.

Futuyma, D. J. and M. Slatkin. 1983. *Coevolution*. Sunderland, MA: Sinauer Associates. I believe this to be the best of several books on coevolution. Twenty-three contributors present a very wide-ranging series of examples (not confined to the tropics). Somewhat technical.

Golley, F. B., and E. Medina, eds. 1975. *Tropical ecological systems: Trends in terrestrial and aquatic research*. New York: Springer-Verlag. Consists of twenty-five technical papers by various authors ranging from rain forest topics to savanna and mangrove ecology.

Golley, F. B., ed. 1983. *Tropical rain forest ecosystems: Structure and function*. Amsterdam: Elsevier Scientific Pub. A collection of technical reviews by the experts covering many aspects of tropical ecology.

Halle, F., R.A.A. Oldeman, and P. B. Tomlinson. *Tropical trees and forests: An architectural analysis*. New York: Springer-Verlag. Asks and attempts to answer the question, Why do tropical trees look as they do? Quite theoretical and rather technical.

Janzen, D. H. 1975. *Ecology of plants in the tropics*. London: Edward Arnold. Excellent though brief account of plant adaptations in the tropics. Emphasizes allelopathy, seed dispersal, pollination ecology. For the general reader.

Janzen, D. H., ed. 1983. *Costa Rican natural history*. Chicago: Univ. of Chicago Press. This is the best general reference book currently available on Neotropical ecology. It contains brief life histories of selected species from all major taxons plus very well written overviews of the ecology of climate, geography, agriculture, plants, mammals, reptiles and amphibians, and arthropods. In all, 174 contributors. It covers only Costa Rica and is by no means all-inclusive, but it is very information packed. If you read one additional book on tropical ecology, it should be this one. Well illustrated.

Larsen, K. and L. B. Holm-Nielsen, eds. 1979. *Tropical botany*. New York: Academic Press. Good collection of technical papers on all aspects of tropical botany. Separate chapters on different regions.

Leigh, E. G., Jr., A. S. Rand, and D. M. Windsor, eds. 1982. *The ecology of a tropical forest: Seasonal rhythms and long-term changes*. Washington, DC: Smithsonian Institution. This volume contains thirty-two papers that, if you read them all, will certainly convince you of the importance of seasonal variability in the tropics.

The papers deal mostly with Barro Colorado Island in Panama. Several of these papers were discussed in this book.

Longman, K. A., and J. Jenik. 1974. *Tropical forest and its environment.* London: Longman. A very brief general introduction to the structure of tropical forests.

Loveless, A. R. 1983. *Principles of plant biology for the tropics.* London: Longman. This is basically a botany text specifically slanted to the tropics.

MacArthur. R. H. 1972. *Geographical ecology*: Patterns in the distribution of species. New York: Harper & Row. A fine introduction to tropical ecology through a comparison with the temperate zone. Excellent discussion of species diversity. Technical.

Moser, D. 1975. *Central American jungles.* New York: Time-Life. Readable, almost chatty, account of rain forest and jungle. Good for gaining an overall impression of the tropics. Excellent illustrations.

Moynihan, M. 1976. *The new world primates.* Princeton, NJ: Princeton Univ. Press. Excellent brief survey of Neotropical primates but does not include much about leaf palatability.

Myers, N., for the Committee on Research Priorities in Tropical Biology of the National Research Council. 1980. *Conversion of tropical moist forests.* Washington, DC: National Academy of Science. Emphasizes loss of tropical forests region by region.

Myers, N. 1984. *The primary source: Tropical forests and our future.* New York: W. W. Norton. This is an overview of conservation issues involving the world's tropical areas. Excellent and concise presentations of the many complex issues, mostly revolving around the consequences of deforestation. For the general reader.

Perry, D. 1986. *Life above the jungle floor.* New York: Simon and Schuster. Layperson's account of Perry's unique research in the rain forest canopy.

Prance, G. T., ed. 1982. *Biological diversification in the tropics.* New York: Columbia Univ. Press. This volume consists of thirty-six technical papers examining the refugia theory for diversification in the tropics.

Rappole, J. H., E. S. Morton, T. E. Lovejoy III, and J. L. Ruos, eds. 1983. *Nearctic avian migrants in the Neotropics.* Washington, DC: U.S. Department of the Interior, Fish and Wildlife Service. Essential reference for anyone interested in the subject of migrants on their wintering grounds. Excellent summary essay, data tables, maps, and very comprehensive references.

Richards, P. W. 1952. *The tropical rain forest.* Cambridge: Cambridge Univ. Press. Dated now but nonetheless still useful. Discusses soils and vegetaion but no animal ecology. Written mostly for the professional botanist.

Richards, P. 1970. *The Life of the jungle.* New York: McGraw-Hill. A layperson's guide to what rain forests look like and how they function. Abundant illustrations.

Snow, D. W. 1976. *The web of adaptation*. New York: Demeter Press-Quadrangle. Superb brief discussions of sexual selection and frugivory among tropical bird species. Emphasizes manakins, cotingas, and oilbirds, all from the author's and his wife's work.

Snow, D. W. 1982. *The cotingas: Bellbirds, umbrellabirds, and other species*. Ithaca, NY: Cornell Univ. Press. A lavishly illustrated book summarizing in detail Snow's work on this group of birds. Many examples of sexual selection.

Sterling, T. 1973. *The amazon*. Amsterdam: Time-Life International. Another in a series of layperson's books on major gobal ecosystems. Not much hard-core information but enjoyable. Well illustrated.

Sutton, S. L., ed. 1983. *Tropical rain forest: Ecology and management*. Palo Alto, CA: Blackwell Scientific. Discusses forest structure and function, and plant-animal interactions, through conservation issues.

Tomlinson, P. B., and M. H. Zimmerman., eds. 1978. *Tropical trees as living systems*. Cambridge: Cambridge Univ. Press. A collection of technical papers from a symposium. Deals mostly with structure and function of trees.

UNESCO/UNEP/FAO. 1978. *Tropical forest ecosystems: A state of knowledge report*. Paris: UNESCO, available through Unipub, NY. General but technical survey of biology of tropics with sociological aspects. Includes case studies.

Wallace, A. R. 1895. *Natural selection and tropical nature*. London: Macmillan. This delightful book contains vivid descriptions, in Victorian prose, of Wallace's experiences in Amazonia.

Walter, H. 1971. *Ecology of tropical and subtropical vegetation*. New York: Van Nostrand Reinhold. A standard technical reference on the characteristics and classification of tropical and subtropical ecosystems.

Waterton, C. [1825] 1983. *Wanderings in South America*. Reprint. London: Century Publishing. A very entertaining narrative by a rather eccentric but perceptive explorer.

SCIENTIFIC NAMES MENTIONED
IN THE TEXT

GENERAL INDEX

Illustrations and diagrams are indicated by boldface type.

coevolution (*cont.*)
ant plants, 163–65; and chiropterophily, 161–63; of coral reefs, 361; of fungus gardens and leafcutting ants, 165–68; and pollination, 158–61. *See also* frugivory; mycorrhizae
coffee, 104
Coley, Phyllis D., 183, 188, 191
Colombia, 30, 48, 75, 114, 117, 141, 150, 316, 335
Colwell, Robert K., 56, 228
competition, 120, 126–29; diffuse, 120
conch, queen, 355
condor, Andean, 24, **25**
Connell, Joseph, 154
Connor, Edward F., 153
conservation: of tropics, 363–76
conservation data centers (CDCs), 373–74
Conservation International (CI), 374
convergence, intertropical, 4
convergent evolution, 223, 255
Cook, Captain James, 106
copra, 108
coral: boulder, 356; brain, 360; elkhorn, 356, 360; finger, 360; gorgonian (sea fan), 361; mountain star, 360; sheet, 360; soft, 360; staghorn, 360; starlet, 360; stinging, 360, 361
coral reef, 28, 155, 357–62; and coevolution, 361; comparison of, with rain forest, 361–62; diversity of, 358, 361–62; and green algae, 360; kinds of, 359; productivity of, 361; zonation of, 359–60
Corcovado National Park, 200
Cordilleras, Mexican, 12
Corn, 91, 100, 101, 102, 370
Costa Rica, 13, 30, 34, 45, 48, 56, 72, 78, 83, 97, 101, 104, 106, 127, 128, 135, 156, 162, 165, 166, 169, 172, 173, 174,

187, 192, 193, 194, 196, 200, 202, 228, 234, 283, 291, 306, 315, 326, 328, 374
cotinga (cotingidae), 172, 211, 218, 236–37, 341; barethroated bellbird, 242; bearded bellbird, 170, 242–44, **243**; bellbirds, 236; cocks-of-the-rock, 236, 246, 249; fruitcrows, 236; fruiteaters, 236; Guianan cock-of-the-rock, 238–41, **239**, 246, 250; lovely cotinga, 172; pihas, 236; purple-throated fruit-crow, 170; purpletufts, 236; screaming piha, 236, 241–42; three-wattled bellbird, 242, **243**, 244; umbrellabirds, 236; white bellbird, 242
crab, 44; fiddler, 351; land hermit, 28, 350, 351, **352**; mangrove tree, 350
crocodilian, 310, 313–14; American crocodile, 314; black cayman, 314; cayman, 232, 287, 300, 310, 314; Morellet's crocodile, **314**; speckled cayman, 18, 314
crops, tropical, 102–8
crossbill, red, 341
crypsis. *See* cryptic coloration
cryptic coloration, 114–16
ctenosaur. *See* lizard
Cuba, 226, 275, 335
curassow (Cracidae, Galliformes), 214–16; great curassow, **215**; helmeted curassow, 215; razor-billed curassow, **216**
Cuzco, 14, 21

Darwin, Charles, 3, 16, 30, 47, 109–12, 119, 123, 133–34, 159, 238–41, 275, 296, 325–26, 343, 359
Davis, William E., Jr., 171, 274
decomposition, 68